"Damn the Torpedoes"

CONTRIBUTIONS TO NAVAL HISTORY . . . NO. 4

"Damn the Torpedoes":
A Short History of U.S. Naval Mine Countermeasures, 1777–1991

Tamara Moser Melia

Naval Historical Center
Department of the Navy
Washington, D.C. 1991

Library of Congress Cataloging-in-Publication Data

Melia, Tamara Moser, 1955-
    Damn the torpedoes : a short history of U.S. naval mine
countermeasures, 1777–1991 / Tamara Moser Melia.
        p. cm. — (Contributions to naval history ; no. 4)
    Includes bibliographical references and index.
    ISBN 0-945274-07-6
    1. Mines, Submarine—United States. 2. Mine sweepers—
United States. I. Title. II. Series.
V856.5.U6M45 1991   91-15918
359.9'383'0973—dc20

For sale by the U.S. Government Printing Office
Superintendent of Documents, Mail Stop: SSOP, Washington, DC 20402-9328

From *A Treatise on Coast-Defence . . .* by Viktor von Scheliha.

Secretary of the Navy's
Advisory Committee on Naval History
(As of 1 January 1991)

William D. Wilkinson, *Chairman*

CAPT Edward L. Beach, USN (Retired)

David R. Bender

John C. Dann

RADM Russell W. Gorman, USNR (Retired)

Richard L. Joutras

VADM William P. Lawrence, USN (Retired)

Vera D. Mann

Ambassador J. William Middendorf II

VADM Gerald E. Miller, USN (Retired)

Clark G. Reynolds

Daniel F. Stella

Betty M. Unterberger

Contributions to Naval History Series

The Author

Tamara Moser Melia earned her doctorate in historical studies from Southern Illinois University at Carbondale, where she assisted in editing six volumes of the *Papers of Ulysses S. Grant*. She has been employed as a historian for the U.S. Navy at the Naval Historical Center since 1982, working on a variety of early and contemporary naval topics. In addition to her present assignment as the Center's Fleet Liaison, tasked with documenting current naval operations, she also is an adjunct professor on the faculty of Georgetown University and an active member of the Surface Navy Association. Her next project is a special study of mine warfare in the Persian Gulf since 1980.

Foreword

Damn the Torpedoes recounts the United States Navy's longstanding efforts to counter enemy sea mines. The author demonstrates that interest and capabilities in this special area of naval warfare often waned throughout the course of naval history. When the reality of hostile mines materialized, however, and it became clear that these relatively inexpensive and often unsophisticated weapons posed a deadly threat to America's use of the sea, the Navy rose to meet the formidable challenge.

The author faced her own challenge in preparing this lucid account of a complex story. A large body of information exists on America's association with mine warfare. That experience dates back to the American Revolution and continues today in the mine clearance campaign underway off the coast of Kuwait. Dr. Melia undertook the difficult task of selecting the important strands from the voluminous record of the Navy's operations to counter minefields. Her writing reveals an immediate understanding of this subject gained through experience with the Navy's current mine countermeasures operations, including three recent tours with fleet units operating in the Persian Gulf. We thank the members of the mine warfare community for the superb opportunities they provided to Dr. Melia to learn about their activities on a first-hand basis.

The assistance of other individuals also needs special acknowledgment. In 1989 several historians and mine warfare officers from outside the Naval Historical Center provided valuable comments on an earlier version of this manuscript. These reviewers included Professor William R. Braisted of the University of Texas; Captain Paul L. Gruendl, USN (Retired); Captain Donald E. Hihn, USN (Retired); Rear Admiral Charles F. Horne III, USN (Retired); Captain Bailey Liipfert, USN; and Professor Jon T. Sumida of the University of Maryland.

In preparing her study for publication, Dr. Melia worked closely with Captain Steven U. Ramsdell, USN, Director of the Naval Historical Center's Naval Aviation History and Publications Division. Captain Ramsdell offered excellent historical and editorial advice. The highly competent staff of *Naval Aviation News*, led by Lieutenant Commander Richard R. Burgess, USN, designed and typeset this book. Finally, special recognition must go to Sandra Doyle, the Naval Historical Center's senior editor, for her dedicated professionalism in overseeing the conversion of a manuscript into a first-class publication.

Despite the support provided by these and other individuals, the opinions and conclusions expressed in this history are solely those of Dr. Melia. They do not reflect the views of the Department of the Navy or of any other agency of the U. S. Government.

Dean C. Allard
Director of Naval History

Preface

In July 1987 supertanker SS *Bridgeton*, under the protection of U.S. Navy warships in the Persian Gulf, was damaged by a simple contact mine of 1907 design. Responding quickly, the Navy hurriedly dispatched all available air, surface, and undersea mine countermeasures assets to provide protection for American-flagged ships transitting the gulf. Most Americans seemed surprised to discover that these assets consisted of a few 30-year-old surface ships, most from the Naval Reserve Force; aging helicopters that suffered greatly in the gritty gulf environment; and mine countermeasures specialists recalled from retirement. As political cartoonists lampooned the Navy for its seeming unpreparedness to counter mine warfare in the gulf, analysts began to probe the history of the Navy's mine warfare program in general and its mine countermeasures efforts in particular.

I observed these events unfolding over the course of the summer and fall of 1987 from my desk in Washington, where I was employed as a Navy Department historian. In October, Dr. Ronald H. Spector, then the Director of Naval History, asked me to produce a short, unclassified narrative history of mine countermeasures. What he originally envisioned, a brief synopsis with large appendixes of original documents, proved impossible to produce because of large gaps in extant original documentation and a lack of previous focused study on the subject in the available literature. The assignment soon evolved into a short, strictly unclassified narrative of episodes in the history of mine countermeasures, to be written solely from secondary sources.

I made several discoveries along the way. The first was that the subject deserved far more than a brief, episodic survey. The second was that there was no shortage of unclassified, secondary material on the subject. Indeed, sorting through the sheer mass of extant material, much of which was unindexed, haphazardly organized, and of varying quality and treatment, brought me to a third realization: to relate accurately the history of the study and application of mine countermeasures over the past two hundred years, I would need access to original source material, classified or unclassified, and to as many of the people involved in the science and art of countering mine threats as I could possibly find.

As this manuscript grew to over double the intended size, the only part of the original plan that remained intact was the book's focus. Although mine countermeasures cannot be divorced completely from the broader subject of mine warfare, the primary focus of this book is still the history of the evolution of mine countermeasures within the U.S. Navy. Due to the limits imposed on this brief study, it was not possible to include every relevant technology, plan, policy, asset, or operation in complete detail. Neither was it possible to name all the many people who, at one time or another, did much to advance mine countermeasures within the Navy. While I have made no attempt to fix blame or attach glory to any individual or circumstance, I have attempted to trace those people and events that affected mine countermeasures and to discover why events unfolded as they did.

During my search for material, I found masses of documents, files, reports, studies, and correspondence on mine warfare and mine countermeasures in all periods of history in the National Archives; the Naval Historical Center's Operational Archives; and the technical library of the Fleet and Mine Warfare Training Center in Charleston, South Carolina, which holds the archives of Commander Mine Warfare Command. While there is no shortage of original material available for additional studies on mine warfare and mine countermeasures described in these pages, researchers should note that the select bibliography ending this study does not by any means indicate all sources consulted and reviewed but is a listing of primarily unclassified, secondary studies useful to this subject. Indeed, those interested in writing further on mine countermeasures should be encouraged by the number of subjects for which the bibliography has few listings.

The purpose of this book then is not to provide a definitive interpretation of any one aspect of or operation involving U.S. Navy mine countermeasures activity or to make recommendations for the future but, rather, to put the whole subject into historical perspective. My ulterior motive is to encourage further writing by those who are interested in the subject or who have experienced life in mined waters. Several of these individuals have materially assisted me by describing, correcting, suggesting, and redirecting my efforts to understand their experiences and to put them into the proper context. Although their efforts are responsible for markedly increasing the quality of this manuscript, all views, errors of fact and interpretation are my own and do not necessarily reflect those of the Department of the Navy. All corrections and additions to the story are welcome, both for inclusion in the holdings of the Naval Historical Center and for future revisions of this book.

Tamara Moser Melia

Acknowledgments

This study would not have been possible without the assistance of literally hundreds of active duty, reserve, retired, and civilian surface, aviation and undersea experts who patiently explained their specialties and the intricacies of mine countermeasures. I deeply appreciate the opportunities I have had to work with the fleet, and particularly want to recognize the men of *Adroit*, *Avenger*, *Fearless*, *Iowa*, *Leader*, *Tripoli*, mobile sea barge *Wimbrown 7*, HM–14, HMS *Brecon*, HMS *Hecla*, BMS *Loire*, HrMs *Zierikzee*, Mine Squadron 2, the many EOD detachments, and the successive staffs of the Persian Gulf MCM group commands and the Mine Warfare Command (MINEWARCOM) for taking the time and effort to educate me. I am also greatly indebted to many officers on the ships and on the staffs of the allied minesweeping task groups operating in the gulf in 1991, particularly Capitaine de Vaisseau Ph. Grandjean and Capitaine de Frigate Philippe Convert of the French Navy; Commander (s.g.) Bob Cuypers of the Belgian Navy; Commander Klaus Saltzwedel of the German Navy; Commander Mike Nixon, Lieutenant Commander Mike Croome-Carroll and Lieutenant Commander John R. Staveley of the Royal Navy; and Lieutenant Commander Robert Gwalchmai of the Royal Canadian Navy.

I am particularly grateful for the advice and information provided along the way by Vice Admirals John W. Nyquist and Joseph S. Donnell; Rear Admirals Robert C. Jones, Donald A. Dyer, Raymond M. Walsh, and William R. McGowen; Captains Charles J. Smith, Paul L. Gruendl, Cyrus R. Christensen, Donald E. Hihn, Jerry B. Manley, Frank J. Lugo, Richard W. Holly, Paul X. Rinn, Larry R. Seaquist, Robert L. Ellis, Bernard J. Smith, Richard S. Watkins, R. A. Bowling, James Chandler, Paul A. Cassiman, Anthony J. Kibble, Bill Hewett, Mike Rogers, David W. Vail, and Bruce McEwen; Commanders Tom Beatty, Daniel G. Powell, Robert S. Rawls, Richard L. Schreadley, Jim Aaron, James Y. Gaskins, Greg Greetis, Peter E. Toennies, Bruce E. Dunscombe, Dave Stewart, Leonard P. Wales, R. B. Jones, and Craig E. Vance; Lieutenant Commanders Steven A. Carden, Brad Goforth, Jake Ross, Rod Scott, Mike Steussy, J. D. Cope, Elliot Powell, Jr., Paul S. Holmes, Francis X. Pelosi, Justin M. Sherin, Jr., Kelly M. Fisher, T. Scott Wetter, and Mark B. Yarosh; Lieutenants David E. Halladay, Terry Miller, Daniel F. Redmond, Paul F. Burkey, Rick Muldoon, James M. Speary, and Frank Thorpe; Lieutenant (jg) Geoffrey B. Bickford; Master Chief Bobby Scott and Boatswain's Mate Wilfred Patnaude of *Iowa*; Chiefs James G. Richeson, M. K. Thulis, and W. J. Walters; Petty Officer Second Class Joe Bartlett, Dave Belton, Jeffrey Bray, Harold Langley, Charles Myers, Philip

K. Lundeberg, Jack Combs, Patrick A. Yates, William Yohpe, Scott C. Carson, Tom McCallum, and Donald Zweifel.

Rear Admirals Byron E. Tobin and William W. Mathis, two successive commanders of MINEWARCOM during the preparation of this book, provided me with valuable information and corrected many of my errors. Several of their staff officers also read all or part of different drafts of the manuscript for accuracy and security review. Of these I would especially like to thank Captains Franklin G. West, Jr., and Richard Fromholtz and Lieutenant Commanders Richard Leinster and Robert McMeeken for their comments and suggestions. Lieutenant Commanders Steve Nerheim and Joseph Tenaglia of Surface Force, U.S. Atlantic Fleet, gave a careful and critical analysis of the manuscript, as did Dr. Ray Widmayer at OP–374 and Jan Paul Hope, Assistant Director of Naval Construction, Office of the Assistant Secretary of the Navy (Shipbuilding and Logistics). Captain Spence Johnson patiently read several drafts of the manuscript and made important contributions and suggestions. I am particularly grateful to Commander Robert B. O'Donnell of the CNO Executive Panel for critical editing of the final draft.

Several historians and textual reviewers also substantially added to the value of this volume. I am especially grateful to Virgil Carrington "Pat" Jones, whose epic three-volume *Civil War at Sea* broke the code on Farragut's minehunting efforts in 1962. In addition to those previously mentioned, I also appreciate the work of Drs. William S. Dudley, Thomas C. Hone, Lynne Dunn, Edward J. Marolda, William Braisted, Allan Millett, and William N. Still; Rear Admiral Charles F. Horne III; Captain Richard Wyttenbach; Commander Robert McCabe; Major Charles Melson, USMC; Scott Truver; and A. T. Mollegen, Jr., who reviewed one or more drafts of the manuscript and suggested improvements.

This manuscript was prepared under the supervision of two Directors and two Deputy Directors of Naval History, Drs. Ronald H. Spector and Dean C. Allard, and Captains Charles J. Smith and A. J. Booth, to whom I am grateful for their advice and encouragement. Janice Beatty, Charles R. Haberlein, Kathleen M. Lloyd, John C. Reilly, Jr., Dale Sharrick, Judith Short, and Gary Weir of the Naval Historical Center; Iris Rubio of the Fleet and Mine Warfare Training Center Technical Library; and Richard Von Doenhoff of the National Archives provided valuable documentation above and beyond the call of duty. Captain Steven U. Ramsdell provided me with substantive editing and technical advice and went to the mat for me on several major issues, for which I am truly grateful. Ms. Doyle's assistant, Akio J. Stribling, and the staff of SSR Inc. capably copyedited the manuscript and indexed the book. Charles Cooney is responsible for the excellent graphics, design, and layout; and Joan Frasher typeset the manuscript.

I owe a personal debt to friends and supporters who believed in this project, especially Norman Friedman for information, advice, encouragement, and general mentoring; and to Rear Admirals Anthony A. Less and Raynor A. K. Taylor for unprecedented operational opportunities in the Arabian Gulf in 1988, 1989, and 1991 with both American and allied MCM forces. John Y. Simon, Judy Randall, and Alfred Khoury were there when I needed them. Paul Gruendl deserves more than a generous share of thanks for the documents, photographs, maps, guidance, and corrections he provided. I found additional inspiration in the command histories of the late Captain Felix S. "Hap" Vecchione, an articulate and realistic spokesman for mine warfare, widely believed to have been the Navy's best hope for a mine warfare admiral before his untimely death.

Finally, I owe great thanks to my family, particularly to my husband, Patrick, for his faith, and to my son, Grant (USNA Class of 2012 hopeful), for inspiration.

Contents

Illustrations

Photographs with NH and NR&L numbers are held by the Naval Historical Center, Washington, D.C. 20374–0571; those with 80G, 19N, K, and USN numbers are held in the Still Pictures Branch, National Archives, Washington, DC 20408–0001; and those with DN numbers are from the Department of Defense Still Media Records Center, Washington, DC 20374–1681.

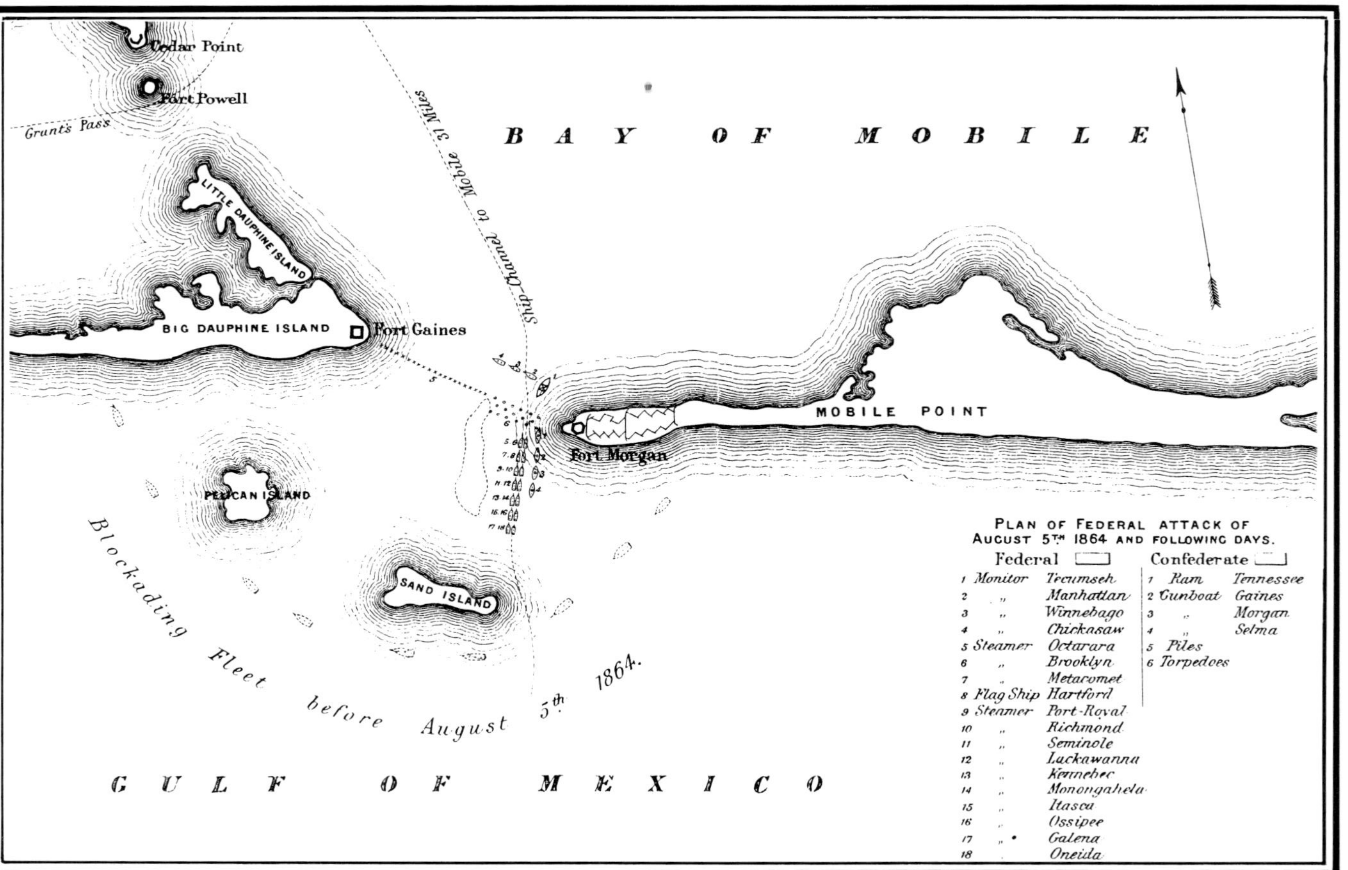

Map of the entrance to Mobile Bay showing the disposition of Farragut's fleet. From *A Treatise on Coast-Defence . . .* by Viktor von Scheliha.

Introduction

Historians often describe their art by using the German terms *historie* and *geschichte*. Historians collect *historie*, or the story of what happened, from raw materials such as memoirs, lessons learned, command histories, letters, recollections, oral histories, traditions, legends, and sea stories. They then question each source, analyze its significance, and try to discover *geschichte*, or what actually happened. Such historical truth cannot be discovered without sustained application and continued analysis of lessons learned, as well as thorough study of the real effects of both failure and success.

In order to examine why the U.S. Navy, with all its capabilities, has not sustained an effective interest in mine countermeasures, we must look beyond legend and, step by step, attempt to discover the lessons learned and unlearned along the way. What we may well find is that lack of mine consciousness and adequate historical perspective has often led the Navy to remember the wrong lessons from its mine warfare experience.

Rear Admiral David Glasgow Farragut's dramatic entrance through the mine line into Mobile Bay in 1864, for example, has become an enduring legend of naval history and an important lesson used in the training of naval officers. The incident is also an object lesson in the history of the Navy's attitude toward the subject of mine countermeasures. The image of Farragut created by the popular press, that of a daring man who risked an unknown mine threat to defeat the enemy and become the most honored officer in the Navy, influenced generations of naval officers. In terms of both Farragut and the mine threat, however, the Navy has remembered the wrong lessons. What actually happened is a lot more interesting than the legend.

Between Mobile's Forts Morgan and Gaines, Confederates had narrowed the deep-water channel approach to the bay with underwater pilings and three staggered rows of approximately 180 moored mines about seventy-five feet apart, leaving a clear passage only under the guns of Fort Morgan. More than two-thirds of these mines were cone-shaped tin Fretwell-Singer mines. These mines, planted in May 1864 by Confederate Army torpedo expert Lieutenant Colonel Viktor von Scheliha, were fired by direct contact between a ship and the mine's cap and trigger device. A few other mines, mainly Brigadier General Gabriel J. Rains's keg-type wooden ones with ultrasensitive primers, had been planted since February. On the bay's floor lay several huge electrically-fired powder tanks that were controlled from shore. Farragut observed Confederate Admiral Franklin Buchanan's men remining the bay daily, noting that "we can see them distinctly when at work."

When the Confederates extended the minefield further east, marking the easternmost point with a red buoy, the open path forced ships under the guns of Fort Morgan; Farragut decided to find a path farther west. While probing the western obstructions in preliminary forays in early July, Farragut's men quietly began pulling up sample mines, often called "torpedoes," and passing them among the fleet for study.[1]

Late Sunday night, on 31 July 1864, as one witness recalled it, Lieutenant John Crittenden Watson, Farragut's flag lieutenant and personal friend, set out on a curious expedition to Mobile Bay. Heavily armed, Watson and his men rowed quietly into the bay with muffled oars. Unobserved by the enemy on this particular occasion, they proceeded with their nightly mission of examining a field of these moored Confederate contact mines blocking much of the entrance to Mobile Bay.[2]

Although this foray through the mine lines is the best documented example of Watson's minehunting activities, surviving ship deck logs and memoirs attest to his repeated attempts to gain information for Farragut on the extent of the mine danger zone. Watson took picket boats out from Farragut's flagship, the sloop of war *Hartford*, on the nights of 30 June, 25 July, and 27 July; on 25 July he had been accompanied on an extended survey by boats from gunboat *Sebago* and sloop *Monongahela*.

Each night Watson and his boat crew methodically worked down the three lines of mines and found that many of the Fretwell-Singers, which were anchored or suspended from buoys about ten feet below the water, had deactivated during the long immersion. Watson's crew drilled holes in the buoys to sink them, removed them, or simply cut them adrift. What Watson found did not surprise Farragut. Earlier in the campaign, in February 1864 at nearby Fort Powell, Farragut had discovered that the percussion mechanisms of the vaunted Fretwell-Singer mines had been deadened by marine worms, allowing Union vessels to cross the mine lines without incident.[3] The similar leaky condition of the mines recovered by Watson at Mobile reassured Farragut that "only a few were very dangerous."[4]

On 28–30 July, side-wheel steamer *Cowslip* towed Watson and his boat crew around the bay, marking safe passage with buoys. Farragut sent Watson out again on 1–3 August with orders to sink the rest of the buoys holding up mines, "as he thinks they support what will otherwise sink," clearing more unmarked paths through the minefield. Watson reconnoitered so close to shore on one expedition that he was able to bring back five deserters from Fort Gaines under the guns of the Confederates' mightiest forts. On 3 August, on his last expedition, Watson was accompanied by pilot Martin Freeman and Farragut's personal secretary, Alexander McKinley, who later recorded their assiduous efforts to sink the mines. Farragut observed Confederate crews laying another ninety mines, probably of the keg variety, on 3–4 August and noted their placement carefully.[5]

In his battle orders Farragut assured his fleet that the mine line had been thoroughly surveyed. "It being understood that there are torpedoes and other obstructions between the buoys, the vessels will take care to pass eastward of the easternmost buoy"; which he knew from Watson's reconnaissance should be "clear of all obstructions." [6]

At high tide on the morning of 5 August Farragut entered the bay. Farragut's officers had persuaded him to allow the sloop *Brooklyn*, hastily rigged with a rudimentary torpedo catcher on the bowsprit, to lead the advance in place of Farragut's flagship, *Hartford*. Contrary to Farragut's orders, monitor *Tecumseh* moved west of the red buoy where it struck and detonated one of the newly placed mines. As *Tecumseh* quickly went down, *Brooklyn* suddenly stopped and backed, stalling the fleet's advance.[7] High in *Hartford*'s rigging Farragut watched *Tecumseh* sink and *Brooklyn* hesitate. From *Hartford*'s poop deck Lieutenant Watson heard the admiral's exchange with *Brooklyn*: "Farragut hailed again and all that could be distinguished of her reply was something about torpedoes. 'Damn the torpedoes!' he instantly shouted, ordering *Hartford*'s captain 'Full speed ahead, Drayton.'" [8]

Farragut then took the lead, heading the ship and the fleet into the minefield that Watson's crews had so industriously surveyed. Sailors later swore that they heard the Fretwell-Singer mines' steel rods snap as the ships pushed over the triple row of torpedoes, but the mines' leakiness and inactive triggers kept them from exploding. As Farragut crossed the minefield, his flag lieutenant "expected every moment to feel the shock of an explosion under the *Hartford* and to find ourselves in the water. In fact, we imagined that we heard some caps explode. . . . and probably no straighter course was ever kept than by these ships in passing over that torpedo field, the furrow made by the *Hartford* being accurately followed." [9]

On 6 August Farragut issued a general order of thanks to his men for their conduct of the previous day. Unwittingly underplaying Watson's efforts while exaggerating his own ignorance of the state of the mine defenses of the bay, Farragut claimed that of the mines he "knew nothing except the exaggerations of the enemy."[10] Some of Farragut's officers later asserted that Watson's minehunting efforts were far more extensive than even Farragut himself knew. "How far he entered the Bay on these occasions in the darkness of the night I doubt if the Admiral knew himself," one man reported, although "he evidently was in close touch with the enemy."[11] Confederates on the scene agreed that however they were discovered, Farragut knew exactly where the live mines were. Lieutenant F. S. Barrett, the Confederate mining officer at Mobile Bay, observed Farragut's approach on 5 August and later stated he believed that "it is evident they were well informed as to the location of the torpedoes we had planted."[12] After the battle Farragut's men took up the obstructing mines at the entrance to the bay and found that most were indeed harmless.[13]

Farragut's daring in the face of enemy mines was real. Before the battle Farragut had decided to enter the bay whatever the obstacles. Confederate mine warfare had, however, been so successful in stalling the Union advance that by 1864 no captain in the Union Navy could afford to ignore the mine threat when attacking a fortified harbor. Watson's efforts no doubt added to the failure rate of the mines, and the knowledge he gave Farragut of the extent and exact parameters of the mine threat influenced Farragut's decision to take the risk. Farragut prepared to battle the mines as carefully as he prepared his vessels for the fight, gaining sufficient information about the condition of the minefield before him to make a dangerous yet measured decision. Farragut did not, as many assert, merely "damn" the mines at Mobile Bay but, rather, assiduously hunted, examined, and disabled them before steaming into the bay. His meticulous approach to the mine threat is a crucial lesson in risk assessment that, unfortunately, most contemporary observers missed.

* * *

Mine warfare is by definition the strategic and tactical use of sea mines and their countermeasures, including all offensive and defensive mining and protection against mines. Mining and mine countermeasures (MCM) are, however, two distinctly different operations.

The primary focus of modern mining operations is to effect sea control, with secondary missions that neutralize or destroy enemy ships by interdicting enemy sea lines of communication, submarine operating areas, and home ports. Offensively, mines attack enemy ships in transit or bottle them up in their own waters; defensively, mines guard national and international waters against enemy intrusion.

Anything actively or passively undertaken to defeat the mine's function of attacking or sinking enemy vessels can be broadly defined as MCM. Traditionally viewed as a defensive measure, in recent years MCM has come to mean something more. Since the Civil War, MCM craft have often been at the forefront of offensive operations, leading strike forces into and out of ports, clearing channels and staging areas in advance of the fleet, and following the MCM force motto, "where the fleet goes, we've been." In a larger sense offensive MCM operations can involve most Navy assets and may require the assistance of land forces as well. Attacks on enemy minelayers, detonation control centers, and manufacturers have successfully proven that the most effective counter to sea mines is to prevent them from being planted. MCM operations today aim to reduce the threat of enemy mining by both offensive and defensive operations against enemy minelaying agents and their supporting facilities.[14]

Mines are composed of different combinations of explosive charges, firing mechanisms, sensors, and housings. They can be described by their position when planted as bottom, moored, previously moored floating, or as drifting mines. However, they are most often and accurately described by their sensor mechanism as contact (requiring contact with the hull of a vessel to detonate), controlled (remotely operated by electrically-generated cables), or influence (generally actuated by a ship's magnetic, acoustic, or pressure "signature," or by a combination of these signatures and other factors).

Contact mines, historically the most commonly used, are generally moored and can be countered by cutting the mooring cables and allowing them to pop to the surface for removal or detonation. Influence mines are most often bottom laid, depending on the target and depth of water. Modern mines are often controlled by ship counters, delay systems, or self-destruct mechanisms, and can be laid by surface ships, aircraft, or submarines. The most sophisticated types of new "smart" mines, which use microcomputers housed in homing torpedoes, can be set to target specific types of enemy ships and to turn themselves off to avoid detection and removal.[15] Before mines can be countered, however, they usually have to be found, classified, and exploited for their intelligence value in order to determine how best to neutralize them.

MCM efforts can be passive or active. Passive measures are defensively designed or used to avoid detonating a mine and are often employed by surface vessels. They include:

1. *Mine watching*—pinpointing areas where mines are laid, usually done by human mine spotters or electronic sensors.

2. *Mine avoidance*—marking mined areas or rerouting of waterborne traffic.

3. *Deperming*—demagnetizing a hull by electrical current, thus nullifying the vessel's magnetic field through periodic external application.

4. *Degaussing*—nullifying a ship's magnetic signature through installation of permanent equipment on board.

5. *Noise reduction*—reducing the likelihood of a vessel actuating a mine by installing noise reduction features or procedures during shipbuilding.

6. *Ship-protection devices*—using a range of devices to protect a ship from mines, from the early use of nets and booms at anchor and underway, bow watches, and early torpedo catchers and rakes (ancestors of the World War II paravane) to experimental mine-avoidance sonar.[16]

Active MCM are usually offensively designed and used by trained MCM forces to locate and neutralize mines without harm to the vessel or to fool mines into detonating on a false target. They include:

1. *Minehunting*—searching waters for mines, from the early use of small boats with searchlights and divers to advanced minehunting sonar.

2. *Mine disposal*—physically neutralizing mines, either through gunfire, deactivation, sympathetic explosion, or physical removal or destruction by divers.

3. *Mechanical minesweeping*—using minesweepers, either singly or in pairs, towing wire and cable or chain rigged with buoyed sweep gear to mechanically cut mooring cables, allowing the mines to surface so that they may be neutralized. Mechanical minesweeping is the most common form of MCM.

4. *Influence minesweeping*—creating false signatures by a towed device or combination of devices designed to produce the magnetic, acoustic, pressure, or other influence needed to explode mines at a safe distance from the sweeping vessels. Minesweeping vessels must also be passively protected to prevent or to limit the almost inevitable damage that accompanies minesweeping.

5. *Countermining*—attempting to clear mines through the use of underwater ordnance or explosive charges placed most often by divers, or most recently by remotely operated vehicles (ROV).

6. *Removal*—physically removing mines from waters, accomplished only by divers and at great risk to personnel.

The history of MCM is a history of progress, decline, and resurgence. Regular cycles of peacetime neglect have stalled funding and study of appropriate countermeasures as mine technology has advanced. Wartime exigency has led to the hasty development of effective operational mine countermeasures to meet a specific threat, only to be followed, once that immediate threat was successfully met, by loss of interest in further development of MCM. Given this cyclical interest in MCM, some knowledgeable observers have compared the development of countermeasures technology today to a periodic reinvention of the wheel.[17] This institutionalized failure to sustain the lessons of history has periodically led to embarrassment. Those who believe that Farragut nullified the mine threat at Mobile Bay solely through daring and fortitude may condemn the Navy, and the nation, to remember all the wrong lessons from the history of MCM.

1

A Matter of Efficacy:
Countering Contact Mines
1777–1919

Mine warfare in America began with an individual, David Bushnell. Famed for his experiments with underwater explosives and submersibles during the American Revolution, Bushnell also attempted with little success to attach explosive charges fixed with primitive clockwork primers to the hulls of ships. His best attempts were countered easily by primitive means but not without casualties. In the summer of 1777 Bushnell cabled together a double line of contact mines to attack the British frigate *Cerberus* off the coast of Connecticut. A British prize crew riding alongside *Cerberus* in a captured schooner spotted the mines and hauled the line in, exploding the mines, which killed most of the crew and destroyed the schooner. *Cerberus* was undamaged, however, and the frigate's captain found and destroyed the rest of the mines.[1] Bushnell fared little better when he took on the entire British fleet above Philadelphia on 7 January 1778. The buoyed contact-primed kegs of powder he sent drifting down the Delaware River arrived in daylight and were seen and avoided. Although four British sailors died in an attempt to retrieve one of the kegs, the British ships again escaped harm.[2]

A generation later Robert Fulton also labored to convince major powers in Europe and America to invest in both mines and fired torpedoes. Despite Fulton's successful experimental mining of the British ship *Dorothea*, Nelson's victory at Trafalgar six days later made the point moot. Britain retained control of the seas and had no need of Fulton's mines, which were described as "a mode of war which they who commanded the sea did not want, and which, if successful, would deprive them of it."[3] Fulton later induced Secretary of State James Madison and Secretary of the Navy Robert Smith to fund additional experiments, but his efforts to make the mines sink a ship required so many attempts that nearly everyone grew skeptical.

In 1810 Congress funded a test of Fulton's "harpoon torpedo," a primitive contact mine with a spring-lock firing pin that followed the path of a harpoon tether into a ship. U.S. Navy Commodore John Rodgers and Captain Isaac Chauncey assisted the congressional commission that evaluated the experiment. Ordered by the commission to protect a ship against mine attack by passive means, without employing either guns or men to counter them, Rodgers swathed Lieutenant James Lawrence's brig *Argus* at the New York Navy Yard with a heavy net from the frigate *President*. Fixing the net in place with spars and grapnel hooks, Rodgers weighted it with kentledge—"heavy pieces of metal suspended from the yard arms ready to be dropped into any boat that came beneath them, and scythes fitted to long spars for the purpose of mowing off the heads of any who might be rash enough to get within range of them."[4]

To Fulton's disappointment Rodgers's jury-rigged mine protection device defeated the harpoon torpedoes, and Rodgers ridiculed Fulton's efforts to attack naval vessels with such "cheap contrivances." Fulton acknowledged Rodgers's "ingenuity" in countering the mines but pointed out that such extreme countermeasures certainly proved the importance of his invention: "A system, then only in its infancy, which compelled a hostile vessel to guard herself by such extraordinary means could not fail of becoming a most important mode of warfare." Despite Fulton's argument the congressional commission was impressed by the ease with which Rodgers had foiled the mines. Persuaded that mines could be countered easily, Congress declined to fund further research or to purchase mines.[5]

Fulton's biggest contribution to mine warfare, his development of the first moored contact mine, designed the same year but overlooked by his own generation, ultimately influenced mine design. This mine, a cork-floated copper tube with gunpowder and musket fittings, safely deactivated itself with the passage of time and popped to the surface for removal.

Convinced that his mines would make navies unnecessary, Fulton assiduously campaigned for funding and caused many navalists to oppose his mines on principle. Additionally, his experiments produced one of the most persistent, though unintended, attitudes towards mine warfare. Most American observers agreed with the British that mines would be best used by nations with the weakest navies. Neither American nor foreign navies gave much thought to preparing countermeasures for so unlikely a mode of warfare.[6]

By 1842 Samuel Colt—later of revolver fame—had perfected an electrically-controlled mine detonation system in which the charge was exploded from shore by a human operator at the precise moment of the ship's passage. Colt's experiments with controlled mines attracted much attention in Washington. After President John Tyler and his Cabinet witnessed Colt's destruction of a schooner on the Potomac River in August 1842, Congress appropriated funding for further tests. Although Colt destroyed several ships both at anchor and underway in experiments between 1842 and 1844, he failed to convince Congress or the Navy Department that mines would significantly improve naval warfare.[7]

Development of mine warfare in Europe from 1844 to 1860 influenced American interest in the weapons. The French improved Colt's designs without finding much operational use for the system; similar controlled mines laid against the Danish fleet at Kiel in the Schleswig-Holstein War,

1848–1851, were never tested. Russian production and use of contact-fused mines during the Crimean War, 1854–1856, was the first systematic defensive employment of the weapon. The Russians laid a series of minefields, mixing contact and controlled observation mines in likely allied anchorages at Sevastopol, Sveaborg, and Kronstadt. The simple but clever contact mines severely damaged British ships, and Royal Navy seamen learned quickly to recover and disarm them. Lacking a more systematic approach to the problem presented by mined harbors, the British, however, could do little to counter mines in massive numbers. Two British vessels, *Merlin* and *Firefly*, reconnoitered the Russian minefields near Kronstadt in 1855 and brought back enough information on the threat posed by the mines to cause the British to cancel a planned attack.[8]

Interested in the new technology and tactics used in the Crimean War, the U.S. Army sent three of its best officers to the field as official observers. Army Captain Richard Delafield, an early advocate of defensive harbor mining, inspected the Russian mine system and reported favorably on its use. The standard countermeasure to these early mines remained small guard boats, which were used to patrol the waters for the easily spotted shallow-moored mines, and bow watches on warships, a system successfully employed by the British in 1857–1858 against Chinese mines.[9]

The outbreak of civil war in America in the spring of 1861 soon challenged the U.S. Navy's previous hesitancy to use mine warfare. Clever improvisation by Confederate naval personnel using land mines underwater, and subsequent development and use of more advanced water mines in nearly every southern river and harbor, forced the Union Navy to devise countermeasures and to employ a few mines of their own. Most naval officers did so unwillingly, however, and agreed with Admiral Farragut's reluctance to employ mines in retaliation. "Torpedoes are not so agreeable when used on both sides, therefore, I have reluctantly brought myself to it," he wrote Secretary of the Navy Gideon Welles. Expressing a traditional view of mine warfare, he continued, "I have always deemed it unworthy of a chivalrous nation, but it does not do to give your enemy such a decided superiority over you."[10]

Equally reluctant to fund mine research, Welles left most ordnance matters to his ship captains and allowed mine countermeasures to be developed and applied ad hoc throughout the fleet.[11] Each captain designed his own protection devices, if any, and most officers found the presence of mines more tedious than hazardous, at least at first. As one officer rashly remarked early in the war, "All contact torpedoes are liable to be removed and overcome by ordinary ingenuity, if it is allowed full exercise by uninterrupted operations."[12] Naval ordnance expert Captain John A. Dahlgren noted that "so much has been said in ridicule of torpedoes that very little precautions are deemed necessary."[13] Such Union attitudes changed only with recurring demonstrations of the effectiveness of such primitive devices, particularly in increasingly frequent Confederate guerrilla operations. Some Union commanders became unwilling to risk their ships in mined waters without direct orders and often found rumors of Confederate mining in the eastern rivers sufficient reason for inactivity.

The Confederacy, on the other hand, actively funded mine warfare as an inexpensive alternative to traditional naval defense for a nation without much of a navy. Confederate inventors Matthew Fontaine Maury, Beverly Kennon, Hunter Davidson, and Gabriel J. Rains experimented with torpedoes and earned renown for their firing mechanisms. Maury's particular interest in electrically-fired mines led to the creation of the Confederate Submarine Battery Service, which developed and detonated most of the controlled mines planted during the Civil War.

Davidson relieved Maury in command of that service, and Maury spent most of the war perfecting his electrical mines in English laboratories. As his work progressed, Maury's mines became more lethal. In October 1862 the Confederate Congress funded a separate Torpedo Bureau. This army unit was headed by Rains, who had been experimenting with land mines since the Seminole Wars in Florida in 1840 and had already mined the tributaries of the James River with experimental contact mines. A third group of Confederates in secret service placed the more insidious "coal torpedo" and similar package bombs aboard Union vessels.[14]

The increasing effectiveness of Civil War mines and mining methods affected the study and use of mines both defensively and offensively until well into the twentieth century. As the war progressed, both Union and Confederate naval officers experimented extensively with offensive mining, particularly with drifting mines and spar torpedoes projecting from the bows of specially constructed torpedo boats and ironclads. The simplest Confederate contact mines, or even the threat of such mines, remained the most effective method of stopping or redirecting Union ships. Two particular types of Confederate contact mines proved most effective. Rains created a keg-type friction mine, using tarred oak barrels with wood end cones for buoyancy and stability. When the glass and chemical fuse contacted a ship, the chemicals broke into a chamber filled with alcohol and liquid gunpowder and exploded the tightly packed charge of powder housed in the sides of the barrel. The second most widely used Confederate mine was the tin Fretwell-Singer, which worked by spring action rather than a fuse.[15]

Ultimately, officers who understood the methods of firing torpedoes were most successful in developing appropriate active and passive countermeasures. Few Union officers hunted mines as actively as Farragut had at Mobile Bay, but many employed forces of small patrol boats in advance of their river and coastal fleets to search for mines. Most often, paired small boats would tow either end of a chain or cable weighted in the center to catch the cables that anchored the buoyant contact mines either to the river bottom or to a nearby float. Contact with the sweeping chain would either explode the mine harmlessly between the boats or snag it on the mine cables, allowing divers or other small boats to remove or destroy the mine. Contact mines that could not be easily pulled from the water or cut from their moorings could be countered by a single bullet making a hole large enough to flood the powder chamber, rendering the mine inert without removing it from the river. If mines were hauled ashore, they could also be bored with hand tools. Controlled mines were countered by advance river shore patrols that searched the riverbanks for electrical cables or detonation centers and dragged for the cables with grappling hooks trailed from small boats. To foul floating mines, some captains, like John Rodgers had done before them, standardized a passive system of logs and nets to swath their ships at anchor.[16]

The first mines found by Union forces were pulled out of the Potomac River in July 1861 by Commander S. C. Rowan, commanding the sloop *Pawnee*. Mines on the eastern rivers quickly became such a significant threat that in early 1862 Welles commissioned shipbuilder John Ericsson to design something to "remove or destroy" underwater obstructions and mines. Ericsson, however, remained diverted from the project by his ship designs and his plans were never carried out.[17]

Almost as soon as Union naval vessels appeared in force on southern rivers,

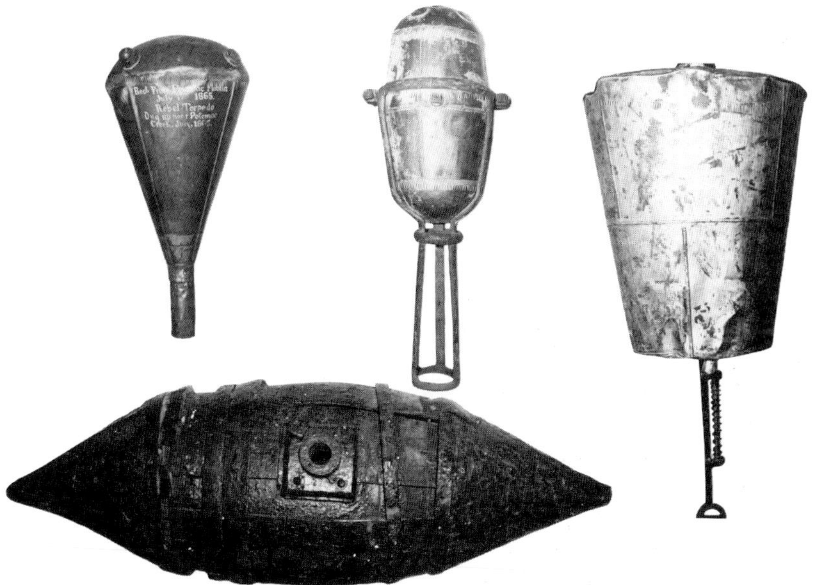

Courtesy U.S. Naval Institute

Confederate mines (*clockwise from lower left*): Keg type, Brooks type, Spar torpedo, Fretwell-Singer type.

NH 53872

The Civil War gunboat *Genesee* is swathed with antitorpedo netting for mine protection. Engraving from *Harpers Weekly*, 27 May 1865.

Confederate mines claimed their first victims. The first major Union ship loss to a mine, however, was not until December 1862. During that month a Union naval mine clearance expedition down the Yazoo River in Mississippi encountered difficulty when boats dragging for electrical wires and contact mines ahead of *Cairo* suddenly opened fire on floating mines. Mistaking the firing as a Confederate attack on his advance, Lieutenant Commander Thomas O. Selfridge ordered *Cairo* ahead to assist the boats. Discovering his error, he stopped and sent his small boats out to assist in recovering the mines. As Selfridge turned his drifting ship aside to resume course, *Cairo* came under Confederate gunfire and maneuvered over a remote-controlled "demi-john torpedo." Confederates on shore fired the mine with two electrically-activated friction mechanisms, and *Cairo* swiftly sank.[18]

The loss of *Cairo* immediately encouraged the Confederacy to accelerate development and use of mines and the Union to improve minehunting and sweeping efforts. Impressed by the loss of *Cairo*, Acting Rear Admiral David Dixon Porter, to whose flotilla *Cairo* was attached, immediately began directing the development of ship protection measures for the advance ships in his command. One of his officers, Mississippi Marine Brigade Colonel Charles R. Ellet, designed an effective "torpedo catcher," also known as a "torpedo rake" or "devil," a system of wrought-iron hooks and logs welded to a spar on the ship's bow-sprit at keel level that extended 20 to 30 feet forward to scoop up mines in the fleet's path. Encouraged by Ellet's success, Porter had torpedo catchers built for many of his ships on the western rivers. Private inventors began sending him their designs, and ships throughout the Union Navy soon sprouted torpedo catchers from their bows.[19]

Porter extended his personal interest in countering mines throughout his fleet and often sought ingenious means to foil Confederate minelayers. When preparing for expeditions upriver from Vicksburg, Mississippi, in July 1863,

Porter ordered that temporary torpedo catchers be made for every ship in his command, personally describing the method to be used and encouraging suggested improvements from the fleet. In addition to conducting minesweeping and controlled cable hunting on the rivers, Porter actively hunted entire minefields, plotting mine positions to reveal the configuration of the field. Vigilance also had an effect on halting minelaying. After the fall of Vicksburg, Porter so thoroughly patrolled the Mississippi that Confederates could no longer mine the river effectively. Not all of Porter's attempted countermeasures, however, succeeded. As he retreated from the disastrous Red River expedition in the spring of 1864, he sent two small cutters sweeping in advance of his prize ironclad *Eastport*. Although the boats detonated two mines, *Eastport* hit a third and went down, forcing Porter to destroy her to avoid having the ship refloated and captured by the Confederacy.[20]

Few Union commanders had as much success at countering the Confederate mine threat as did Porter and Farragut. In the attack by his ironclads on fortified Charleston harbor, Rear Admiral Samuel F. DuPont similarly rigged his ships with torpedo catchers fabricated from old scrap iron, spars, nets, and grapnels. Monitor *Weehawken* showed notable success in clearing some of the mines in her path. When pummelled by Confederate shore artillery, however, she could do little to penetrate the defenses of the city. "The ghosts of rebel torpedoes have paralyzed the efficiency of the fleet," one observer wrote, "and the sight of large beer barrels floating in the harbor

The Union monitor *Saugus* on the James River displays her torpedo rake or catcher.

. . . added terror to overwhelming fear. . . . The torpedo phantom has proved too powerful to be overcome." [21]

Hoping that more active measures would yield better results, DuPont's successor, Rear Admiral John A. Dahlgren, ordered Ensign Benjamin H. Porter to minehunting duty. For several nights in September 1863 Porter, like Lieutenant Watson at Mobile Bay the following year, sought to discover if mines had indeed been laid between Forts Sumter and Moultrie and if other obstructions were present. Determined Confederate opposition, however, prevented him from discovering much.

Dahlgren fretted so over his lack of knowledge of the underwater conditions at Charleston that he considered sending divers down to investigate as battle raged overhead. While preparing to support Major General William T. Sherman's army in capturing that city, Dahlgren fabricated torpedo catchers by rigging two 50-foot logs across each ship's bow and dangling nets to scoop up any floating torpedoes. As the fleet moved toward the city on 15 January 1865, two monitors sailed ahead with a convoy of small boats sweeping with grapnels. Despite Dahlgren's precautions, however, the monitor *Patapsco*

U.S. Navy Photograph

The Farragut Window in the Naval Academy Chapel heroically portrays the admiral "damning" the torpedoes from the rigging of his flagship, *Hartford*, at Mobile Bay.

Courtesy U.S. Naval Institute

struck a torpedo and sank that night. There was little more Dahlgren could do with his limited resources; the mines at Charleston proved so effective against the Union fleet that no one ever successfully countered them.[22]

Watson's efforts did not end the problem of mines in Mobile Bay. Pressed by the encroaching Federals, Confederates hurriedly mined all the water approaches to the city after the battle even as Farragut's men worked to clear mines from the channel entrance. Some mines exploded while being landed on shore, leaving five of Farragut's men dead and eleven wounded. Farragut also lost one small boat while sweeping. Confederate torpedo expert Lieutenant Colonel von Scheliha and his replacement, Lieutenant J. T. E. Andrews, continued to mine all water and land approaches to Mobile for the rest of the war, and by March 1865 eight vessels of the West Gulf Blockading Squadron had been sunk by mines despite continuous countermining efforts. By the war's end more than 150 mines had been retrieved from the waters around Mobile. Most of this work was done in the steamer *Metacomet* by Commander Pierce Crosby, whose thorough sweeping with nets garnered him twenty-one floating mines.[23]

As Confederate mines improved, individual Union commanders increased their ability to avoid them, but the cost was high. Despite the experience at Charleston, Union countermeasures proved so effective that Confederate inventors developed supposedly unsweepable secondary mines. Attached to the primary explosive at a distance, these mines were designed to explode under the minesweeping ship upon the successful sweeping of the main mine casing. The Brooks mines, with such a secondary exploding device, made sweeping rivers difficult, if not impossible. Maury also continued to improve his controlled electric mine batteries at his laboratory in England and smuggled some improved equipment into the Confederacy for use in the eastern rivers.[24]

As the Union Navy moved inland into the South by river, the Confederates increased their use of advanced controlled mines, particularly in the James River approaches to Richmond. It was here that Major General Benjamin F. Butler brought his Army vessels under the command of Navy Acting Rear Admiral Samuel Phillips Lee, combining forces in an effort to capture Drewry's Bluff in May 1864. Warned of controlled mines in the James by former slaves, Lee dragged for mine cables with all his vessels and small boats and sent shore patrols along the riverbank.

So carefully and slowly did Lee's advance proceed that waiting Confederates were able to spot the movement and transfer all their detonating equipment to the opposite bank. Steamer *Commodore Jones*, leading the vanguard, moved forward carefully, surrounded by a busy guard of rowboats sweeping with grapnels. Despite Lee's precautions, *Commodore Jones* backed over a torpedo fired from onshore by Confederate Submarine Battery Service operators and was blown to pieces. Operators of the battery captured by Union forces after this incident initially refused to identify other mined spots in the river. When Navy Lieutenant Commander Homer C. Blake put one of them in the bow of his steamer *Eutaw* and proceeded into the minefield, however, the operator reconsidered and disclosed the location of the other mines.[25]

When Lieutenant General Ulysses S. Grant sent a combined operation up the Roanoke River in December 1864, the bows of the Navy's supporting gunboats were swathed in nets to protect them from floating mines. The nets, however, provided little protection against contact mines. Near Jamesville, North Carolina, gunboat *Otsego* detonated contact mines and sank. Sweeping near the wreck, surviving sailors found "a perfect nest of torpedoes." While transferring the wounded from *Otsego*, tug *Bazely* hit another mine and went down. Six more mines were discovered around her, and two others were caught in *Otsego*'s net. The small squadron moved forward slowly, using row boats to drag for more mines. In this manner the squadron found several more mines eight miles upstream and another twenty-one lined up in rows blocking the river.[26]

Throughout the war Porter continued to look for novel ways to counter Confederate mines. While commanding the naval forces moving up the James River in the last months of the war, the admiral used lines of fish nets to combat floating mines, deployed his minesweeping boats in diagonal formation to increase sweep effectiveness, and constantly patrolled the river to foil Confederate remining attempts. By April 1865 Porter had so thoroughly cleared the James River that he felt safe ferrying President Lincoln to Richmond by water.[27]

Fifty ships, four-fifths of them Union vessels, were crippled or sunk by underwater mines during the Civil War. Ironically, most Confederate ships lost were victims of their own mines.[28] Continuous Union developments in improving torpedo catchers, spars and nets, and calcium- or magnesium-powered searchlights for night hunting of both mines and offensive torpedo boats were the products of individual officers. Fleet captains received markedly little mine countermeasures support from the Navy Department, which neither emulated Confederate institutional support for mine warfare, nor disseminated any lessons learned, nor established an operational minesweeping fleet.[29] At the war's end development of U.S. naval mine countermeasures remained, as it was in the beginning, dependent upon the interest of individual naval officers.

World observation of the success of defensive mining by the Confederacy made mine warfare appealing to cost-conscious nations seeking economical national defenses. Established navies, however, still considered mines to be weapons for inferior navies, and few considered countering the threat as the key to naval superiority. The Danes and the Austrians defended their harbors with minefields in the 1860s, as did Paraguay in the 1866–1868 war against

NH 51932

A Confederate, electrically-fired, controlled mine explodes close aboard *Commodore Barney* on the James River. Engraving from *The Soldier in Our Civil War*, II:229.

Brazil, the Argentine Republic, and Uruguay. After Brazil lost an ironclad to a simple contact mine while forcing Paraguayan batteries, its allies simply evaded further mine damage by passing at high tide. In 1870 the Prussians advertised defensive mining of all their harbors to keep the French at bay, using dummy mines when real ones could not be procured fast enough. Both the Russians and the Turks defensively mined their waters during the Russo-Turkish War of 1877–1878, during which the Turks laid controlled mines in the Dardanelles but never operationally tested them.[30]

Most growing navies in the late nineteenth century found their interest in offensive underwater ordnance centered around development of mine delivery systems. These included torpedo boats and submarines, which were tested operationally in the American Civil War, and self-propelled torpedoes. Efforts to perfect offensive torpedo delivery systems would consistently overshadow development of mine warfare and mine countermeasures in Europe and America until the turn of the century. While Americans were consumed with internal problems at the end of 1860, the British developed the Whitehead Torpedo, the first truly self-propelled torpedo, and spent the next two decades perfecting it. The British Admiralty also set up the Torpedo Committee in 1873 to develop mine warfare and funded a comprehensive mine program. Royal Navy countermining, actively practiced as early as 1876, successfully used electrically-fired line charges of underwater explosives to sympathetically detonate enemy mines. Countermining became the standard British countermeasure of the late nineteenth century. The Royal Navy also countered electrical mines by sweeping for cables with explosive-laden grapnels and protected their newest battleships against mines and torpedoes with nets.[31]

The U.S. Navy of the late nineteenth century, constrained by scarce peacetime resources and internally factionalized by divisions between line and staff officers, established no similar comprehensive mine warfare program and left mine matters in the hands of interested individual officers. Retaining his wartime interest in mine warfare and determined to prepare for incipient war with Great Britain over its assistance to the former Confederacy, Admiral Porter sought to keep mine warfare alive in at least one small portion of the U.S. Navy. In wartime Porter had been fascinated by the possibilities of underwater ordnance. In peace he continued to experiment, introducing the subject to midshipmen at the Naval Academy while serving as superintendent from 1865 to 1869. During his short tenure as military advisor and assistant to the secretary at Navy Department headquarters in 1869, Porter used his vast power to launch a torpedo station and school under the sponsorship of the Bureau of Ordnance for the study of all underwater ordnance and countermeasures.

Housed at the wartime home of the U.S. Naval Academy on Goat Island at Newport, Rhode Island, and under the command of a succession of Porter's former subordinates, Porter's "Torpedo Corps," as it came to be called was a home for secret naval testing and development of both defensive and offensive systems. In the first year of operation the corps built mines and experimental torpedo boats and provided all Navy vessels with mines for offensive testing afloat. With Porter a regular visitor and the corps' nominal protector in the Navy Department, the school received sufficient funds to begin teaching courses to officers and to design and test the remotely piloted torpedo boat *Nina*. At the same time U.S. Army engineers established an experimental mine station for harbor defense at Willett's Point, New York.[32]

Funding for underwater ordnance remained limited throughout the last half of the nineteenth century, and what funds there were generally went to self-propelled torpedo development, with mine warfare a competitive stepchild. Still, Porter and his captains kept interest in mines and mine countermeasures alive in the Torpedo Corps, if nowhere else. By 1872 the Board of Visitors inspecting the corps recommended expansion of the program, suggesting that research and testing of mines and torpedoes should not outpace research on appropriate countermeasures for ship protection.[33]

But Porter remained frustrated with the Navy Department's unwillingness to fund research in mine and torpedo countermeasures. "I am convinced," Porter wrote Secretary of the Navy George M. Robeson in 1873, "that proper attention will not be given to this subject until special instructions are issued from the Department." He also urged the secretary to add more courses at the Torpedo Station, particularly in countermeasures, maintaining that "there is no course of instruction whatever for defense against torpedoes." He encouraged the secretary to increase appropriations for the Torpedo Corps and to foster experimentation to counter all foreign mines and torpedoes.

Calling for the secretary to follow the examples of the British and of the former Confederacy and make mine warfare a priority within the Navy, Porter suggested a wide range of reforms in mine warfare training and fleet organization. Specifically, Porter recommended the separation of the Torpedo Corps from the bureau and its reidentification as a separate defensive naval force, appropriation of sufficient funding for development of a wide range of mine warfare programs, and appointment of a rear admiral to direct mine warfare training and preparedness. Unfortunately, Porter's plea for reevaluation of the Navy's mine warfare program went unanswered, and his Torpedo Corps remained a component of the Bureau of Ordnance.[34]

Failing to impress the Navy Department with the importance of mine warfare, Porter did what he could to advance the study. Stressing that mines not protected by warships or fortifications could be easily swept using existing methods, Porter had better success obtaining funding for offensive measures. From 1869 to 1876 the Torpedo School trained 153 officers in the fundamentals of mine warfare, including the use of electric lights and fast guard boats to foil both mines and torpedo attack boats. Limited mine testing also took place at the Washington Navy Yard ordnance facilities and at the Naval Academy.[35]

Captain Kidder R. Breese, Porter's former fleet captain, took command of the Torpedo Station in the late 1870s and dedicated most of his first year to countermeasures research and application, particularly in the use of electric lighting, mine clearance, and countermining, which was designed to ease passage in mined harbors. With Breese's interest and the application of countermining, which caught the interest of ordnance specialists, countermeasures advanced within the school's curriculum. Captain F. M. Ramsay, the station's next commander, continued Breese's studies in countermeasures; by 1881 the Secretary of the Navy reported that "there are various methods in vogue for the destruction and removal of hostile torpedoes, and these appliances are being constantly perfected."[36]

Captain Thomas O. Selfridge, former captain of *Cairo*, took over the station in 1881 and began working to develop an integrated system for defense against both mines and self-propelled torpedoes. His was the first systematic attempt within the U.S. Navy to develop a mine countermeasures (MCM) program. In addition to increasing the officers' studies in the minutiae of mine construction, arming, and use, Selfridge extended the course of instruction to eighteen months and included courses in torpedo defense systems, diving, and more advanced countermining and channel-clearing systems. He also instituted practical exercises, hands-on work in countermining, underwater use of electric lighting to spot mines, and ship protection. By 1884 the Navy had ordered advanced electric "torpedo search lights" for its new cruisers, and Selfridge began intensive testing of the lighting system's effectiveness and possible future applications for mine countermeasures.[37]

When Selfridge's tenure at the Torpedo Station ended in 1885, the practical advancement of MCM research also ended. Competition arose between the Torpedo Station and the nearby Naval War College, founded in 1884, over the college's mandate for control of the theoretical study of "success in war." The Torpedo School retained its function of practical training in the manufacture and use of ordnance, but "theoretical" courses in chemistry and electricity were reduced or eliminated. Thereafter, mine production was the primary mission of the station, and MCM activities were limited to practice in the basics of countermining, minehunting, and simple minesweeping.[38]

Despite wartime experience in the United States and the fledgling effort to establish professional expertise in MCM after the Civil War, most naval authorities continued to believe that mines could be easily swept or countermined. By the 1880s the world focus of underwater ordnance research was on the improvement of offensive, automotive torpedoes, and attention turned away from traditional mine warfare. The application of the gyroscope and steam propulsion to torpedoes greatly increased their accuracy, just as Austrian methods of automatic mine depth settings made accurate sea mining possible. In the 1879–1880 War of the Pacific between Chile, Peru, and Bolivia, Peruvian and Bolivian vessels suffered great damage from Chilean spar torpedoes, bomb-rigged vessels, and other moving mine delivery vehicles. By 1882 most navies had a regular torpedo corps that practiced both minelaying and jousting with torpedo boats; concerted countermeasures against submarine mines, however, remained an afterthought. Individual American naval officers closely watched foreign naval experiments with countermining, diving teams, and target practice with underwater objects, but the Navy Department as a whole did little to encourage investment in new technology.[39]

In his 1886 annual report to the Secretary of the Navy, Admiral Porter took the Navy and Congress to task for failing to fund both mine warfare and mine countermeasures, but with little effect. By the following year the Torpedo School curriculum was reduced to only one combined course on countermeasures against both mines and torpedoes and seven practical exercise sessions in mine clearance. In the following years MCM courses remained rote exercises as torpedoes and torpedo countermeasures became the focus of regular experimentation in the United States and abroad. Outside of the Torpedo Station little was done with mines or MCM until an operational need developed.[40]

Real progress in mine warfare and MCM matters in the U.S. Navy throughout the nineteenth century remained the products of the efforts of interested individuals. The Navy Department did not neglect all mine matters after the Civil War, but declining budgets, emerging technology, and lack of strong institutional interest relegated the study to a few men buried in the Bureau of Ordnance.

U.S. support of the Cuban revolution against Spain in the 1890s made diplomatic relations between the two countries sensitive at best. When the battleship *Maine* exploded and sank in Havana harbor on 15 February 1898 with the loss of 266 men, most Americans believed sensationalist press reports that Spain had mined or torpedoed *Maine*. When a U.S. naval court of inquiry declared on 21 March that the explosion was caused by a mine, most Americans reacted to the findings of the court with strong support for war.[41]

Ironically, despite the popular belief that the powerful *Maine* had been destroyed by a mine, there was little interest in developing or using new methods to counter mines or in the growth of an MCM force to meet the weapon that had supposedly devastated a prime example of the modern steel Navy. The only expression of a resurgence in interest in MCM was the preparation of stockpiles of countermines. Using existing resources, the Torpedo Station at Newport quickly designed and assembled electrically-controlled countermines for the fleet, completing the first order of forty mines for the North Atlantic Fleet within ten days.[42]

As Rear Admiral George Dewey prepared for war with Spain, he drew upon some wrongly remembered lessons of his Civil War experience. While serving under his hero, Farragut, at New Orleans in 1862 and at Port Hudson in 1863, Dewey gained experience in running strong land batteries at night. That his ship, *Mississippi*, had been lost in the latter attempt seemed to him merely one of the risks of battle. Dewey had not been with Farragut at Mobile Bay and knew nothing of his careful minehunting efforts. He always believed that Farragut, his acknowledged role model in tactics and risk-taking, had indeed "damned" the torpedoes.[43]

Less than a week after America declared war on Spain, Dewey sought to destroy the Spanish fleet in Manila Bay. Although he had reliable reports from official sources that Manila Bay had been mined, Dewey lacked sound intelligence on the type of mines and their exact placement. He would later explain that he dismissed the mine threat as a "specious bluff" on the basis that the channel's depth, Spain's unfamiliarity with minelaying, the concerted Spanish advertising campaign of the minelaying danger, and the rapid deterioration of mine materials in tropical waters would nullify the likelihood of mines stopping the advance. For Dewey, the advantage of gaining the bay outweighed the risk of the mines. He also recalled a similar incident in 1882 when an Italian mine expert broke a lengthy blockade of the Suez Canal by accurately assessing Egyptian incompetence in minelaying.[44]

On the night of 30 April 1898 Dewey ran the passage into the bay. Like Farragut, Dewey led the advance. When his fleet was halfway past the rumored mine line off Corregidor, the Spanish shore batteries opened fire. The Spanish artillery made such powerful splashes near the ships that the crew mistook them for mines and cried, "Remember the *Maine!*" Nonetheless, the fleet successfully entered the bay without detonating any mines and defeated the Spanish fleet. The only mines fired in the two days of action were controlled mines purposely exploded by the Spanish to clear a path through the mine lines for their own ships to maneuver.[45]

Dewey was both lucky and a sound judge of Spanish incompetence. Charts captured at Cavite on 2 May 1898 noted the placement of the controlled mines in the bay awaiting the detonator cables that had not yet arrived from Spain. In addition, when Dewey sent cruisers *Raleigh* and *Baltimore* to sweep the channel for mines, they discovered that the Spanish had indeed laid several powerful contact mines in the deep channel; their inexperience, however, had led them to simply dump the mines overboard, where they had sunk to the bottom, eighty feet below the ships' keels. Nonetheless, American journalists loudly touted Dewey's passing of an extensive and effective mine danger at Manila Bay, adding Dewey to the folklore of American history as another successful "damner" of mines.[46]

Cuba threatened to prove more difficult. The British passed on rumors of Spanish mining at Cienfuegos, and the U.S. Navy had ample evidence of Spanish mining of all ports, yet never anticipated a real threat. At Guantanamo Bay, U.S. vessels steamed into the harbor in early June without any apparent concern about contact mines. Battleship *Texas* and cruiser *Marblehead* hit two of the mines and ripped them from their anchors without explosions. For two weeks American ships plowed the harbor grazing defective mines before a thorough minesweeping expedition brought up thirty-five ineffective Spanish contact mines. In late July at Nipe Bay the American forces again successfully ignored the Spanish mine threat and passed over a staggered mine line at the entrance to the bay without damage.[47]

The entrance to the harbor at Santiago, a narrow, tortuous channel difficult to navigate under the best of conditions, was a different story. It was defended by both substantial land batteries and electrically-controlled mines, but the mine threat alone stopped Rear Admiral William T. Sampson from attempting to force the channel and enter the harbor. Remembering the twisted wreckage of *Maine*, Sampson readily heeded exaggerated reports of the number and types of mines planted. Calling for land support to capture the mine control bases to allow his big battleships and cruisers to enter the harbor, Sampson settled on a blockade of the Spanish fleet until cooperating land forces could assist him. Sampson made no attempt to countermine the mine lines or to send in small advance parties either to survey the mine positions or to storm the mine detonation positions. He did attempt to bottle up the Spanish fleet by sending in an expedition to sink a ship in the channel narrows. Spanish defenders exploded most of the controlled mines in the harbor to stop the vessel, however, and it came to rest without blocking the Spanish fleet.[48]

The Army forces besieging Santiago planned to attack the harbor's land batteries, allowing the naval vessels to enter the harbor, clear away the controlled mines, and drag for regular contact mines. When Army progress slowed, Sampson considered countermining the bay with forty American countermines stored at Guantanamo, but the admiral concluded that "this work, which is unfamiliar to us, will require considerable time. It is not so much the loss of the men as it is the loss of ships which has until now deterred me from making a direct attack upon the ships within the port." [49] On 3 July the Spanish fleet saved him the trouble by leaving the harbor in an attempt to run his blockade. The subsequent destruction of the Spanish fleet at

NH 002076

The Spanish laid these contact mines in the entrance to Santiago harbor during the Spanish-American War.

Santiago virtually ended the war that ceded Puerto Rico, Guam, and the Philippines to the United States and brought independence to Cuba.

The brief Spanish-American War did little to change contemporary notions about mine warfare. Spanish mines had not been a major impediment to the U.S. fleet, and so few naval officers knew anything about mine warfare that little comment was made about Sampson's failure to scout the minefield at Santiago. At home, American harbors were defensively mined, but the mines stopped only friendly merchant vessels. Few Americans felt the effects of mine warfare in their own waters.[50] Ignoring the mine threat and trusting in Spanish incompetence, the U.S. Navy had the good fortune to escape mine damage and attributed its success to complete U.S. superiority in mine warfare.

After its successful actions at Manila Bay and Santiago, the United States was perceived as an emerging naval power. Although Alfred Thayer Mahan and other proponents of a strong Navy urged Americans to overlook the ease of the victory over Spain and build a powerful fighting fleet, their arguments reiterated traditional attitudes toward mine warfare and MCM, denying their importance on the principle that naval dominance by surface ships made mines unnecessary. Mahan emphasized the historic tradition of sea mines as the weapons of weaker, inferior naval powers, rather than of great powers with command of the seas. Those who followed Mahanian theory often denied the effectiveness of mining and MCM.[51]

Not everyone shared this viewpoint. One notable dissenter was Admiral of the Navy George Dewey. As a result of the influence of the officers of the wartime Strategy Board, in 1900 the Secretary of the Navy appointed a General Board of naval officers with Dewey at its head to act as his senior naval advisors. Perhaps reflecting Dewey's experience, the General Board recommended that MCM testing and training be made part of all regular naval drills. In addition, the board urged that "naval defense mines and mining outfits be prepared and supplied to all battleships and cruisers in such quantities as may seem desirable, and that mining and countermining be made a part of the course of instruction at the Torpedo Station, and also a part of the routine drills on board the vessels having mines on board." The Bureau of Ordnance endorsed the board's recommendations and ordered the new *Maine* class battleships to carry mines and MCM equipment.[52]

However, the available MCM methods had not progressed much since the Civil War. The standard countermeasure to mines in peaceful waters remained small boat minesweeping and grappling for controlled mine cables; the experts recommended advanced controlled countermining while under hostile fire, in keeping with the Navy's focus on weapons rather than on countermeasures as the easiest and most popular method. Small launches towed by gunboats or steamers at full speed would make multiple drops of either single or parallel lines of mines. These mines, armed with 500 pounds

George Dewey "damned" the torpedoes when he ran the mine line into Manila Bay in 1898, but the reality of the mine threat later convinced him to support mine warfare and MCM training for the fleet.

NH 50564

of gun cotton (nitrocellulose), would be electrically command-detonated at a distance from the minefield to encourage sympathetic countermining nearby.[53] As long as mines remained a threat only in shallow waters and harbors, the Navy relied on simple measures that were largely untested in actual wartime operations.

The Russo-Japanese War of 1904 quickly challenged existing concepts of mine warfare as a defensive tactic useful only in shallow waters. Both the Russians, who were seeking to expand into the Far East, and the Japanese, who opposed them, used independently moored contact mines planted in the open sea where small minesweeping boats could not navigate. Hoping to dislodge the Russians from their stronghold at Port Arthur in early 1904, the Japanese mined the waters off the port with sea mines, masking their exact movements, and then lured the Russian fleet out of the harbor on a wild chase. Ignoring the mine danger, Russian Admiral Stephan Makarov in his battleship *Petropavlovsk* led his ships on the return to port right through the Japanese minefield. The admiral and his ship were lost to two mines, and a second battleship was heavily damaged. The Russian minelayer *Amur* retaliated by mining the waters outside the harbor, which resulted in the loss of two first-class Japanese battleships and nearly changed the naval balance in Russia's favor.

An astonished western world watched as minelaying and minesweeping vessels battled each other for control of the sea off Port Arthur. The Japanese, moving in to beseige the port, were forced to begin minesweeping operations to protect their ships. Using tugs with sweep cable armed with sharp cutters to slice through the mines' mooring cables, the Japanese began methodically sweeping paths through the harbor. The Russians quickly learned to foil Japanese minesweepers by quietly moving the buoys marking the cleared paths. This strategy ultimately proved to their disadvantage, for when the Russian fleet was forced to retreat from Port Arthur, it had to pass through its own minefield. Primitive Russian minesweeping efforts, which consisted of paired tugs dragging cables, proved ineffective in sweeping their own mines, and the fleet suffered many casualties. After closely observing the progress of this war, most nations with navies began to explore the possibilities of mining the open sea, thus reviving international interest in minelaying and MCM.[54]

Disturbed by the mine warfare aspects of the Russo-Japanese conflict, some American officers who had scoffed at the Torpedo Corps soon sought to acquaint themselves with current tactics. The regular course of instruction at the Newport Torpedo Station was quickly lengthened from three to six months. The twenty-three Navy students of the 1905 class, along with fifty-two Marine Corps officers and sergeants undertook countermining training that was more advanced than any the Navy had required for decades. That same year the Bureau of Ordnance issued its first pamphlet describing "countermine outfits" for ships and encouraged their use in fleet exercises.[55]

The British quickly reappraised their mine warfare situation and went much further to prepare for operational mine warfare. Abandoning countermining as their primary MCM method, they built up a mine supply and converted existing gunboats to minesweepers. By 1906 they also began testing more reliable minesweeping equipment. Purchasing six fishing trawlers, they converted them to active duty minesweepers between 1908 and 1910. To supplement this regular minesweeping force, the British also recruited private fishing trawlers as reserve sweeps and trained their crews one week annually for pay, a system that many American naval officers urged their service to adopt. These reserve fishermen became the backbone of British minesweeping forces. The heavy winches and cables used to haul up fishing nets were easily adapted to sweeping by fishermen who required little additional training to sweep mines. Furthermore, the British began streaming protective sweep gear from the bows of minesweepers, much as the Union Navy had done in the Civil War, and they developed snags and sweep-evading devices to make sweeping of their own minefields more difficult.[56]

As a result of the Russo-Japanese War, which had littered the Western Pacific with floating mines (live contact mines that had broken free of their moorings), many nations became interested in restricting mine warfare internationally. Signatories to the "Convention Relative to the Laying of Automatic Submarine Contact Mines" drafted by the Hague Conference in 1907 agreed that, for the protection of neutral ships, all moored contact mines would be designed to sterilize if they broke free of their moorings. Most important was the article requiring nations laying mines in international waters to make every effort to remove them after hostilities ended. The covenants also allowed the use of drifting mines only if they were armed with a sterilizer that disarmed them within one hour; they also banned unlimited minelaying. Thus, despite revived international interest in the future of offensive mine warfare, these restricting measures led most nations to abandon thought of devoting increasingly scarce dollars to systematic mine warfare research.[57]

Although most nations developed some mines, there was little change in methods of countering them. The French followed the British example, converting eight fishing boats to minesweepers between 1910 and 1913 and building four more in 1914. Russia and Japan retained a small fleet of minesweepers towing conventional cables, and Germany and Austria experimented with cables and fishing nets. Italian minesweepers added a new dimension by towing small contact mines called "Scotti" that exploded when contacting a mine anchor line. The Italians held regular exercises with their minesweepers, and by 1916 groups of these small boats preceded their fleet.[58]

Despite early interest in new mining methods and countermeasures, the U.S. Navy did little to emulate the efforts of other navies. The General Board urged the Navy to increase studies of mining, minehunting, countermining, sweeping, and removal. By October 1912 Dewey, as head of the General Board, realized its suggestions were being implemented only piecemeal by the bureaus and recommended that coordination of all mining and countermeasures be consolidated under a new Office of Mining and Mining Operations, which would be headed by a Navy captain. Dewey fought for the office against strong opposition from the Bureau of Ordnance and the Bureau of Construction and Repair, which controlled mines, countermining, and minesweeping functions. In the end Dewey was forced to settle for the assignment of an officer at the Naval War College to study mining operations. Until October 1913 the Bureau of Ordnance controlled all mine and MCM material. After that time the Bureau of Construction and Repair controlled MCM equipment, and the Bureau of Ordnance retained mines and mining apparatus.[59]

Like Porter before him, Dewey found that arguments favoring centralized control of mine warfare matters were not sufficiently persuasive to overcome the Navy's established organization. The Navy had suffered few operational losses and had not found mines difficult to avoid. Consequently, other developments in naval warfare continued to have higher priority for the Navy's resources.

By 1912 control of mining and MCM in the operating forces were collateral responsibilities of the commanders of torpedo flotillas of each fleet, who outfitted destroyers with minesweeping equipment. "Owing to lack of experience in minesweeping" in the U.S. Navy, American minesweeping procedures were copied directly from a 1912 British minesweeping manual and reissued yearly between 1915 and 1917. The lightweight, small-boat minesweeping gear carried by the destroyers proved insufficient for practice clearance of heavily mined fields, and by 1913 two fleet tugs, *Patapsco* and *Patuxent*, were rigged with heavier gear and added to the torpedo force. Their success in sweeping during exercises led to the addition of tugs to all fleet torpedo flotillas. Further exercises conducted by the Atlantic Fleet Torpedo Flotilla from 1913 to 1915 encouraged the expansion of the minesweeping force through the development of sweep equipment packages for the faster destroyers and torpedo boats to complement the tugs. In March 1914 Rear Admiral Charles J. Badger, Commander in Chief, U.S. Atlantic Fleet, recommended that all Atlantic Fleet destroyers be fitted out for additional duty as minesweepers. Ten tugs ultimately were identified for minesweeping work, six for the Atlantic Fleet and two each for the Pacific and Asiatic Fleets, although only four were in place before 1917. The separate torpedo flotillas, created in mid-1904 in response to the Russo-Japanese War, thereafter consisted of offensive torpedo boats and torpedo boat destroyers.[60]

With the outbreak of war in April 1914 the Germans swiftly mined the coast of England and began to take their toll of British minesweeping vessels. The British in turn laid mines in the English Channel to oppose German minelaying U-boats, only to be forced to sweep some of them to move their own troops. Later that year the British began mining the North Sea and experimenting with deep-moored mines to stop the submarines. With increased German mining of their coasts, the British expanded their fleet of trawlers, assigning them to hunt submarines in addition to their primary duties of minelaying and minesweeping.[61]

When war in Europe began to impinge on American interests, the U.S. Navy prepared for mine warfare and MCM operations as a part of regular naval combat. In the recently created position of Chief of Naval Operations (CNO), an experienced naval officer was given the responsibility of directing the operations of the U.S. fleets. Recent British experience proved that mine warfare readiness would be a part of the CNO's concerns.

To carry mine warfare to the enemy, in 1915 the Navy transferred its four minesweeping tugs from the Atlantic Fleet Torpedo Flotilla to create the first minesweeping fleet organization, the Atlantic Fleet Mining and Minesweeping Division, tasking its minesweepers to serve also as minelayers. Assistant Secretary of the Navy Franklin D. Roosevelt recommended that minesweeping equipment be prepared for a mixed force of minesweeping destroyers, large fleet tugs, and smaller tugs, but his scheme was quickly reduced to outfitting only three additional fleet tugs and identifying U.S. fishing trawlers for possible future employment as minesweepers. Still, working from the foundation laid by the General Board and the active experimentation of the Torpedo Flotilla since 1912, the U.S. Navy of 1915 had a rudimentary minesweeping organization and improved sweep gear that was deemed sufficient for fleet service, and had also identified ships for collateral minesweeping duty.[62]

The British had problems of their own. Hoping to attack Germany from the south, drive a wedge between Turkey and Bulgaria, and open crucial lines of communication to Russian allies, the British attempted to force the narrow Dardanelles. The Turks had mined the straits in 1914 and constantly remined the passage using both drifting mines and staggered rows of controlled mines. The mines were protected by heavy guns and searchlights on the high ground overlooking the straits below. When the Royal Navy planned to force passage of the straits by naval attack alone, it depended on slow British minesweeping trawlers to sweep a field in advance of the fleet. The trawlers, crewed by civilians sweeping at night under heavy fire and blinding searchlights, made several attempts to clear the mines in advance of the attack but succeeded only in becoming excellent targets for the Turkish gunners. Without gunfire support, minesweepers had little chance of clearing the mines. Two smaller motor launches, joined with a 1000-foot wire, also suffered heavy casualties and failed to clear the mines.

After several unsuccessful night attempts and one daylight effort to clear the mines, the allied Anglo-French naval force undertook a full-scale naval assault on 18 March 1915 with British minesweeping trawlers in the lead. The force's failure to pass the straits, with four battleships lost or damaged in the minefield, was a humiliating defeat for the Royal Navy. British Commodore Roger Keyes reacted by outfitting destroyers with minesweeping gear. The paired destroyers swept successfully for three days under heavy fire in April 1915, but mines eventually forced the British to abandon the attempt to pass the Dardanelles by water and led to the equally ill-fated Gallipoli Campaign. Turkish mines and shore batteries kept the troops at bay until November 1918, when minesweepers preceded the British fleet into Constantinople. Smarting under their embarrassment, the Royal Navy made a considerable effort to develop effective MCM vessels for such assaults in the future.[63]

The British mine warfare experience and German U-boat mining off the American coast forced the U.S. Navy to think critically about MCM, or at least about minesweeping. The individual bureaus continued to adapt British minesweeping doctrine and gear, and operational units integrated minesweeping tugs and instruction into practice. In November 1916 the Atlantic Fleet conducted its first large-scale exercise in mine warfare procedures by laying and sweeping two hundred mines off Sandy Hook, New

Jersey. To prepare surface ships to face mines at sea, Atlantic Fleet Destroyer Force doctrine was amended in May 1917 to include general instructions and procedures for destruction of mines at sea. Commanding officers were ordered to avoid risking their ships in sweeping mines and to use paired ships' boats and fleet picket boats to sweep any area in question.[64]

To untested American crews, minesweeping seemed easy. One reservist described sweeping contact mines with paired ships in five-day rotations off the coast of England: "Sweeping for mines, as I saw, required no great technical knowledge. Rather good sense, considerable nerve, simple precaution, knowledge of sea conditions, and the rest was a matter of chance and persistent, tiring work."[65] American minesweepers also served off France in Patrol Force Squadron 4. The squadron's eleven converted American wooden fishing boats, with French single-ship minesweeping gear armed with explosive cutters, swept mines and escorted convoys without any casualties from mines.

Collateral duties filled much of the squadron's time when it was not sweeping mines. One officer recalled that

> we answered nearby calls for help, patrolled suspected areas, listened in frequently at night for the submarines, assisted in the salvage of several wrecks, rescued a few aircraft, did odd towing jobs, assisted in organizing outgoing convoys from Quiberon Bay, and all in all filled in every odd gap in the work of the district.[66]

Coastal minesweeping trawlers proved so useful as multipurpose boats that minesweeping almost became a subordinate duty to patrols, search and rescue missions, and antisubmarine warfare.

British mine specialists worked to develop more effective minesweeping systems to cut down on the number of ships and time required to sweep an area clear. Traditionally, mines were swept by paired vessels steaming in line abreast and towing a sweep chain or wire. Changes in the interval between the ships, however, sometimes caused the sweep to rise too high, passing over the mines or alternatively narrowing the path of the sweep. Observing North Sea fishing boats, which kept their trawl lines spread through the use of two otter-boards, or diverters, the British adapted these otters to minesweeping. Paired ships towed either end of a sweep wire, held below the surface of the water by a depressor, or "kite," that cut the mine's mooring cables. Most often, however, at least one vessel of the pair was at risk of passing through the minefield.

Early in 1919 the British developed the *Oropesa* sweep, named for the ship on which it was first used, which allowed one ship to sweep a path at a consistent preset depth and thus reduced sweep time. The rig was a long cable with cutters attached, the outboard end supported by a float and guided by a system of diverters that set the outboard depth of the wire. After the sweepers

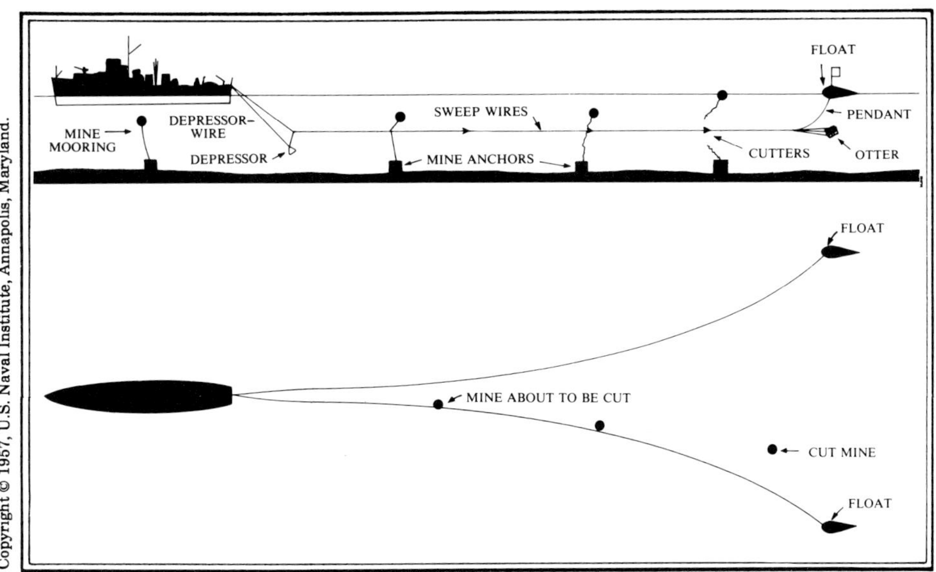

Copyright © 1957, U.S. Naval Institute, Annapolis, Maryland.

Oropesa sweep diagram. From *The Sea War in Korea* by M. Cagle and F. Manson.

made their pass, the mine disposal ships followed, marking the swept field with special "dan buoys" and destroying any moored mines that rose to the surface.[67]

British advances in ship protection inspired American efforts. British captains found that traditional nets used for ship protection underway slowed their ships significantly. To protect ships from contact mines, a Royal Navy commander invented paravanes, torpedo-shaped floats that pulled out sharp wires on either side of a ship's bow at the correct angle to cut mine moorings. A system of hydrostatic pressure valves controlled the paravanes' dual rudders and kept the cables running at a consistent depth.[68] Unable to adapt a similar device then being tested at the model basin of the Washington Navy Yard to meet its needs, the U.S. Navy adopted the British paravane design in May 1917. Three types of paravanes were developed for various ship speeds. Type "M" was designed for slower ships, primarily merchant vessels making up to 16 knots; type "B" for battleships making up to 22 knots; and type "C" for cruisers making up to 28 knots. Because of the high level of secrecy surrounding the paravane system, these devices were often called "otters" as well.[69]

Paravanes protected both warships and private vessels extremely well. By 1918, 2,700 ships worldwide—130 of them U.S. naval vessels—carried paravanes. In addition, the sterns of 150 destroyers were fitted with high-speed mine sweeps—a cross between deep-water paravanes and the *Oropesa* sweep. With these ships ahead of the fleet, cutting a swath 120 yards wide at 25 knots and up to 450 yards at slower speeds, the possibility of countering mines in a swift advance was greatly increased. Repeated examples of success with paravanes encouraged their use. On 7 September 1918 battleship *South Carolina* cut a mine, possibly submarine-laid, with her paravane off New Jersey, saving both herself and battleship *New Hampshire*, which followed in her wake. By late 1918 paravane streaming had become such a normal part of U.S. Navy practice that special paravane courses were added to the curriculum of the Naval Academy.[70]

Mine development throughout the war challenged the traditional minesweeping methods previously thought sufficient. Sensors were added to controlled mines to cue operators ashore to fire them. Counter-countermeasures flourished as navies sought to render their own mines unsweepable, and in successive improvements the Allies put sprocket wheels on mooring cables, allowing sweep wires to pass through the mine mooring cables without cutting them. Other schemes incorporated wire cutters, snags, and explosives on mooring cables to sever the sweep wires. German developments in explosive cutters proved particularly effective, slowing but not stopping British efforts to clear minefields. This family of devices became known as "sweep obstructors."[71]

By August 1918 British scientists had successfully tested a mine more difficult to sweep. This mine, the first magnetic influence mine, rested on the bottom. Its explosion depended on a reading of a ship's magnetic signature,

U.S. Navy Photograph

British and American navies used this type of paravane during World War I.

the permanent magnetic "fingerprint" that steel ships develop unavoidably during construction. Believing them to be unsweepable, the British fitted these mines with devices to render them inert after a predetermined time and used them only rarely.[72]

The greatest menace to Allied ships were the torpedoes and mines employed by German U-boats. To protect U.S. ports, the Navy borrowed the idea of antisubmarine nets from the British early in the war, stringing them across Atlantic harbor entrances where they remained untested throughout the war. German submarines increased their mining of the U.S. Atlantic coast in 1918, and jury-rigged American trawlers swept German contact mines off Thimble Shoals at the mouth of the Chesapeake; Fire Island, New York; Barnegat Light, New Jersey; and as far north as Nantucket Shoals, Massachusetts. Despite these efforts, at least six American ships were lost to mines off the coast. German submarines also laid delayed-rising mines, which reactivated minefields believed clear and heightened the usual danger involved in sweeping mines.

Concurrent development made mines a most effective Allied weapon and accounted for more than 30 percent of German submarine losses in the war.[73] To keep German U-boats out of the North Sea or at least to make passage to Allied waters more difficult, the U.S. Navy Bureau of Ordnance developed the Mark 6 (Mk 6) antenna mine, a relatively cheap, mass-produced mine that proved effective against submarines. These mines required contact between the steel hull of a ship and a copper wire antenna attached to and extending from the mine. Such contact generated an electrical current that fired the mine. Once the mine was developed, a combined British and American mine force in England revived earlier plans for a layered mine barrier across the North Sea. Antenna mines were set in lines of descending depth to form a complete antisubmarine barrier extending 10 feet to 260 feet below the surface, with British contact mines on the wings.[74]

In June 1918 U.S. Navy ordnance specialists began their first offensive mine campaign. By October, 56,611 American Mk 6 mines were laid in thirteen groups, each consisting of rows set 134 miles across at three preset depths. The British planted 16,652 more of their own design. The war ended before the completed barrier could be tested, yet the North Sea Mine Barrage sank at least three U-boats (and possibly three more) during its short existence, and damaged three or four more. More important, it was a psychological barrier to some German submarine crews, at least one of which mutinied rather than pass through the Mine Barrage.[75]

Critics of offensive mine warfare noted that no sooner had the Navy completed planting the barrage than it had to work to clear it, but construction of the first U.S. ships specialized as minesweepers and designed to clear the Mine Barrage had already begun. Attention to specific minesweeper construction began in the fall of 1916, when the Office of the Chief of Naval

Operations (OPNAV) recommended new construction of light draft minesweepers "to be seagoing and with sufficient speed to accompany the Fleet, with power enough to sweep at a speed of ten knots and to be classified as Fleet Sweepers."[76] In response, the General Board developed specifications for minesweepers planned for the 1918 shipbuilding program, and Congress appropriated funds for twelve fleet minesweepers in December 1916.[77]

The Bureau of Construction and Repair's preliminary design for the new fleet minesweeper greatly resembled that of the fleet's minesweeping tugs. It was actually a modification of a 1916 design for a general utility boat, a multipurpose tender or patrol boat. The bureau quickly adapted the design to create a 180-foot, steel-hulled minesweeper drawing 10 feet with a top speed of 14 knots, and the ship went from the drawing board into production in a matter of weeks. Named for birds, fifty-four of these "Bird"-class minesweepers were ordered for delivery within the year. By the time American minelayers had begun planting the North Sea Mine Barrage, these ships were coming out of the yards in time to prepare for MCM operations.[78]

During Allied negotiations in October 1918, the U.S. Navy agreed to clear all American mines laid, or over 80 percent of the mines in an area of 6,000 square miles. New methods were required to sweep these deep-water mines in the turbulent North Sea. U.S. Navy Rear Admiral Joseph Strauss, who had commanded the forces that laid the mines, also commanded the sweep effort. His largely independent U.S. Mine Force, Atlantic Fleet, under British control in minelaying matters, had already given the clearance much thought. Because the U.S. antenna mines were sensitive to the smallest scrap of metal, thin sweep wires could cause them to detonate below a sweeper, setting off chain reactions and possibly countermining whole lines of mines as well. These characteristics also meant that the new steel-hulled "Bird"-class minesweepers could not sweep the North Sea unless a means to foil the steel-sensitive antennas on the Mk 6 mines could be found.[79]

Plans for clearing the North Sea Mine Barrage were not begun until after the fields were planted. Admiral Strauss asked the officers commanding the minelayers for recommendations for quick and safe clearance methods, and they suggested a test sweep with wooden vessels to determine how many mines remained in the field. In December 1918, U.S. Navy Lieutenant Noel Davis, commanding two wooden sailing vessels, *Red Rose* and *Red Fern*, swept a portion of the minefield and immediately detonated a mine. In a string of explosions nearby mines countermined each other, proving their extreme sensitivity. After four passes the two vessels were separated because of bad weather. Operations were postponed until the weather cleared in the spring, at which time Admiral Strauss obtained permission from the Admiralty to borrow sixty British heavy-duty, wooden-hulled steam trawlers to begin sweeping mines.[80]

The loan of the wooden trawlers proved unnecessary. After observing the

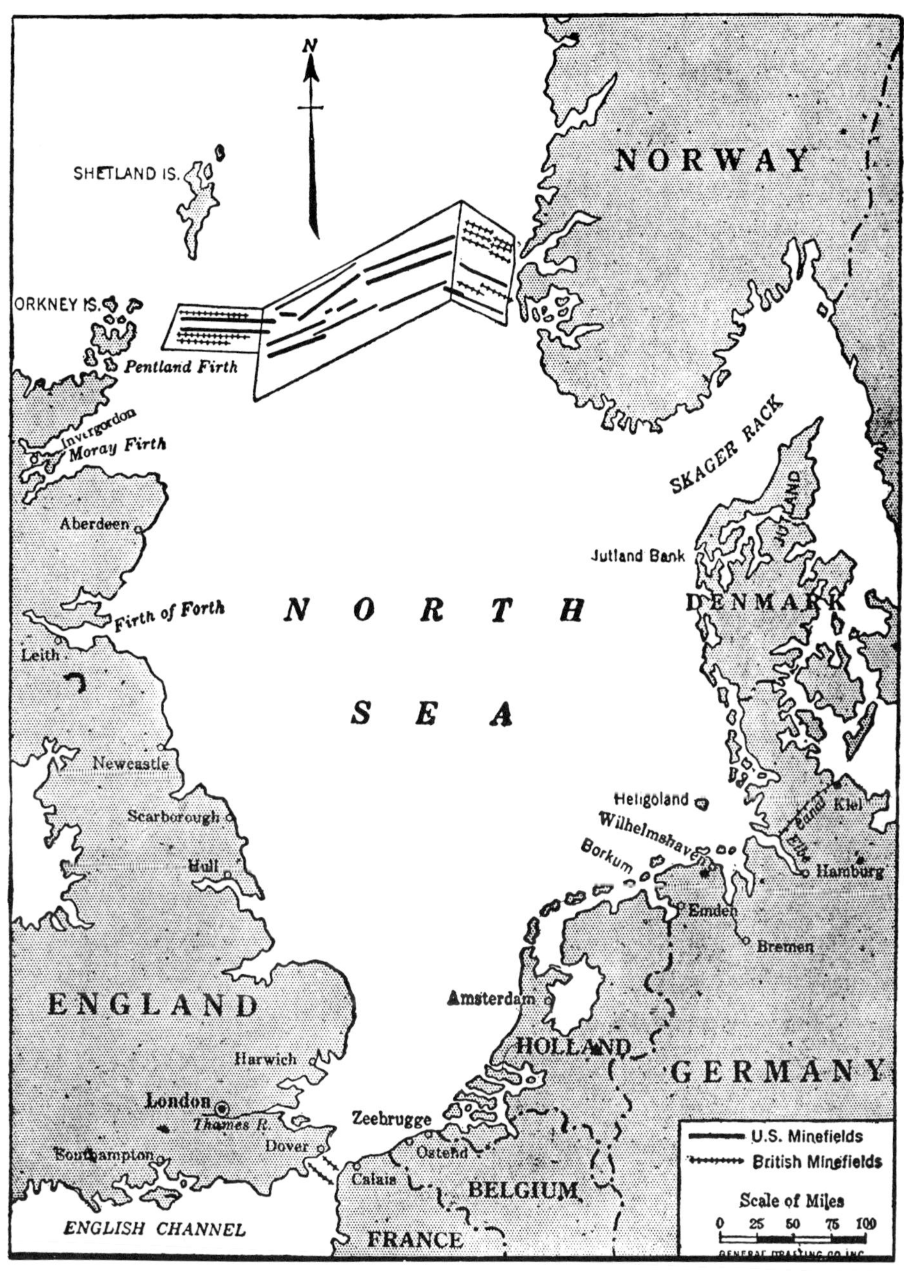

Map of the North Sea Mine Barrage. From *The Victory at Sea* by W. S. Sims.

difficulties involved in attempting to sweep with fragile wooden sailing boats, one of Strauss's staff officers, Naval Reserve Ensign D. A. Nichols, devised an electrical protective device which allowed steel ships to sever the antenna mines' mooring cables without actuating the mines themselves. Nichols's original plan had been to activate the mines by increasing the electrical field in front of the ships. This proved impractical, but by reversing the polarity of the field, mines could be prevented from exploding as the steel ships passed and then be swept by regular mechanical means.

Fitting the old minesweeping tugs *Patapsco* and *Patuxent* with these experimental devices, Admiral Strauss made another test sweep of the minefield on 20 March 1919. The electrical device protected the ships from activating the mines while sweeping, but during the test two problems arose that would plague the entire operation. First, mines would foul the sweep wires and explode before the ships could sink them by gunfire. Second, these explosions would cause mines set at the highest levels to countermine in a chain reaction up to a mile away, often damaging the sweeping ships and carrying away sweep gear.[81]

As British and American forces prepared to clear the mines from the North Sea, British minesweepers encountered severe difficulties in the postwar mine clearance of the Dardanelles. Turkish and German defensive minefields had been intermingled with British offensive types, so it was impossible to determine the different minefield peripheries from the surface. The British sent up observation balloons to locate the different mines and to mark mine paths to be swept. Sweepers plowed closely along the marked lines, sweeping the mines from their moorings as gunboats followed and disposed of the mines. Use of aircraft in spotting the layout of minefields also led the British to use air-dropped countermines and depth charges occasionally; it was the first operational use of air MCM.[82]

During successive sweep operations the American minesweeping force also experimented with different means to offset these problems and to speed up operations in the North Sea. In the first test of the "Bird"-class minesweepers Admiral Strauss experimented with electrical sweep wires to set off the mines, but the electricity altered the ships' magnetic compasses. Returning to traditional mechanical means, Strauss determined that he would need more minesweepers as well as buoy-laying ships to improve navigation while sweeping. Unintentional countermining remained a problem that was as difficult to solve as it was to predict. Admiral Strauss ordered the sweep to begin with the groups of mines set at the lowest depths, thus reducing the danger for the inexperienced sweepers. He also ordered the paired ships to cross the mine lines at right angles in overlapping tracks, a process that quickly proved tedious as well as ineffectual.[83]

As the operation began, the minesweeping force lost large amounts of sweep gear as wires parted and mines fouled or exploded on the kites, requiring

NH 45255

Fleet tugs *Patapsco* (AT–10) and *Patuxent* (AT–11), which joined the torpedo flotilla before World War I, participate in minesweeping operations in the North Sea, 1919.

constant resupply and jury-rigging. Explosive cutters added to the sweep wires in front of the kites lowered such casualties by nearly 50 percent. Although the sweepers were vulnerable to accidental fouling and uncontrollable countermining, the sturdy "Bird"-class minesweepers proved able to withstand all but the closest explosions. By the third operation the minesweeper crews had gained sufficient experience to begin sweeping in compact formation at different depths, which sped up clearance with no appreciable increase in casualties.[84]

Throughout the seven phases of the operation the minesweeping force applied new lessons as learned. Midway through, the buoy-laying squadron successfully detonated the less hazardous lower- and middle-level mines without losing sweep gear by touching the antennas with wires rather than by cutting the mooring cables. To clear the minefield before the fall storm season set in, Strauss's Mine Force used this method to develop a sweep formation for faster clearance. The main body of sweepers then advanced down the lines of mines working at different depths, with the first pair of ships sweeping high to clear mines by engaging the antenna wires and the second and third sets dragging deeper to catch any remaining mooring cables. A subchaser followed to sink mines cut by the sweepers while other ships marked the swept lanes with dan buoys.

Sweeping even in peaceful waters with known minefields required a significant commitment of ships and men, the leadership of creative problem-solvers, and extensive logistical support. As a learning experience, the mine barrage taught the minesweeping forces much about the expensive and dangerous realities of countering mines. Clearance of the Mine Barrage required nearly ten times the assets minelaying had required: 82 ships

NH 471943

Tanager (AM–5) during minesweeping operations in the North Sea, July 1919.

working 17- and 18-hour days for over five months and over 4,000 men, mostly reservists, whose enlistments had been involuntarily extended by the Secretary of the Navy to complete the operation. The sweepers finished clearing the barrage in September with no time to spare. When test sweeping after clearance proved that many mines still remained active, particularly those thought to have been swept in early efforts, a quick but thorough recheck of every field was conducted until no mine exploded. Check sweeping took place despite high September seas.[85] Casualties included two officers and nine enlisted men; one ship sunk; and twenty-three ships damaged.

Complete knowledge of the numbers and placement of the mines and the levels at which they were set had allowed Strauss's officers to apply statistical analysis in developing clearance plans and to plot the number of sweeps necessary for clearance of each field as they went along. Still, when check sweeps revealed no more actuations, thus ending the operation, Strauss's minesweepers could account for only 40 percent of the mines laid. The remainder were assumed to have self-destructed spontaneously in the harsh North Sea weather or by accidental countermining on laying or during sweeping. Events would later prove Strauss's assessment of complete clearance too optimistic, as unswept mines from the North Sea Mine Barrage turned up for many years thereafter. On their return to America Strauss's Mine Force received a hero's welcome with a formal review by the Secretary of the Navy and a ceremonial luncheon in New York City.[86]

Such remembrances were short-lived. Even the assignment of important additional missions to mine force ships did not result in the retention of operational MCM units during peacetime. Within weeks the reserve units disbanded, and the men returned to civilian life. Minesweepers were laid up, and the future of MCM remained in the hands of a few active duty naval officers scattered throughout the bureaus and fleets. The end of war also shelved plans for a new minesweeper design developed by the General Board for fiscal year 1920. Beginning with previous requirements, the board had expanded the minesweeper's mission to include employment in convoy and antisubmarine patrols by beefing up their armament and increasing their speed and maneuverability. In peacetime, however, few officers saw the need for minesweeping ships.

Bureau chiefs and other leaders cited the reluctance of officers to study mine warfare in peacetime as the "reason it has not reached all the development of which it is susceptible." Yet these same leaders were unwilling to make the institutional commitment required to establish mine warfare as a sought-after professional field. Career officers saw no promotion potential or glory in the mine force; consequently, interest in mine warfare after World War I faded quickly. The view that mines could be dealt with through countermeasures contrived ad hoc by adaptation and application of American ingenuity became widespread once again.[87]

Despite clever inventors and interested individuals, U.S. naval MCM throughout the nineteenth and early twentieth centuries developed as a direct

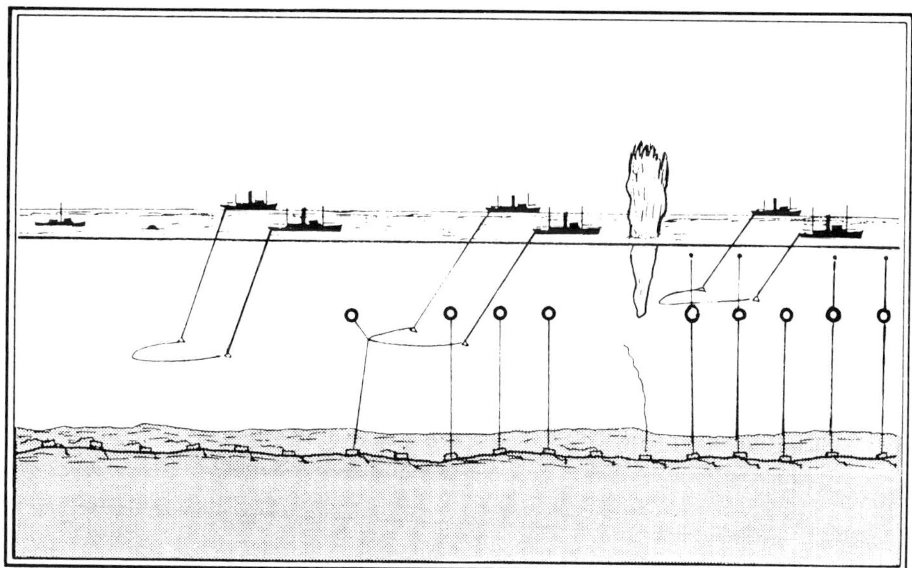

This drawing by U.S. Navy Lieutenant Noel Davis shows paired minesweepers cutting mines at varying levels in the North Sea, followed by a danlaying ship.

response to combat experience. Porter, Dewey, and others who had faced real mine threats in wartime urged peacetime study of the art of countering mines, but serious efforts to do so did not begin until mines began sending ships to the bottom. The actual practice of countering mines changed little from the Civil War until World War I. Ships continued to use bow watches to spot mines, and clever operational officers, American and British alike, developed ship protective devices ranging from elementary torpedo catchers to hydrostatic paravanes.[88]

The development of MCM was also hampered by continuing definition problems. First, MCM was lumped with minelaying as mine warfare, with minelaying usually taking priority over MCM, particularly under the pressures of stingy budgets. Second, development of offensive torpedoes defined mine warfare as strictly defensive and therefore unnecessary to a growing, offensively oriented navy. This approach reinforced contempt for the weapon as a tool of weak or limited powers, easily "damned" by a powerful naval force. Only the operational employment of deep-sea mines against modern capital ships in the Russo-Japanese War indicated that the sea mine was a viable threat for which countermeasures must be prepared, but that experience was soon ignored or forgotten.

For MCM World War I taught an enduring lesson: to operate offensively in mined waters and to defend home waters, a navy must have adequate forces to counter the mine threat and to precede the fighting forces into battle. The British learned this lesson best. British MCM grew in prestige as the Royal Navy faced up to the challenge posed by mines in the postwar world by establishing an active mine warfare school, developing active and reserve minesweeping fleets, and enhancing the promotion potential of MCM officers.

America's World War I experience and its response to that experience greatly differed from that of the British. Relatively few mines disrupted American home waters, and the U.S. Navy experienced no offensive humiliation comparable to that of the British in the Dardanelles. Additionally, success in the North Sea convinced naval leaders that they had adapted suitable and sufficient technology and tactics for using and countering mines. But easy success bred inattentiveness to the future of MCM, and mine warfare remained the province of individuals, rather than a matter of active Navy-wide concern. By the end of World War I, most U.S. naval officers probably agreed with the assessment that minesweeping remained merely "unpleasant work for a naval man, an occupation like that of rat-catching."[89]

2

A New Menace:
The Operational Use of Influence Mines
1919–1945

The British and American navies learned markedly different lessons from their experiences in mine warfare and MCM. Losses to mines in war invigorated British interest in MCM, whereas successful wartime MCM led to American indifference in peacetime. Most American officers credited their wartime MCM efficiency to the quality and training of their reserve force and the inventiveness of their small active duty officer corps. Mines retained their position in American naval tradition as defensive weapons of weaker navies, which the powerful need not fear and American ingenuity could counter.

With peace established, the Navy stored its leftover mines, sent the reservists home, and relegated control of both the weapons and the art to a few interested individuals. Funding for further MCM research and development and a steady shipbuilding program stalled under constrained peacetime budgets and naval arms control agreements. But the stagnation in U.S. naval mine warfare after World War I was due more to continued lack of professional interest within the Navy than to inadequate funding. Officers still avoided prolonged assignment to mine warfare work, believing quite accurately that it was a roadblock to promotion. Without a core of active duty personnel devoted to the study and advancement of MCM matters, MCM leadership was nonexistent, and international interaction remained limited.

Postwar naval leaders did not give mine warfare or MCM the level of priority required to prevent a slide back into their prewar status. Consequently, mine warfare and MCM readiness suffered a similar fate.

In December 1921 the Fleet Base Force was established as a subdivision of the Pacific Fleet and included the provision, at least on paper, of a Base Force, Pacific Fleet, for base defense. Included in this base force was one mine squadron consisting of most of the surviving minesweeping vessels. With the establishment of Battle Fleet, U.S. Fleet, the following year, the base defense minecraft came under the control of the Fleet Base Force, U.S. Fleet, in December 1922. Like its predecessor, the Fleet Base Force consisted primarily of theoretical forces that existed only on paper but which could be added in time of war.[1] By 1928, however, only two "Bird"-class ships remained active as fleet minesweepers. The remaining vessels were stripped of their sweep gear and turned into transports, tugs, and tenders.

When the Navy changed from Battle Force to Battle Fleet organization in 1931, the Fleet Base Force was abolished and renamed Base Force, U.S. Fleet, but still included the same units and structure as its parent organization. Minelayers and minesweepers were combined into a new type command, Minecraft, Battle Force, but no new minesweepers joined the fleet for a decade. Indeed, until 1939 the sole occupant of the mine desk of the Bureau of Construction and Repair, a civilian, simply stored extra sweep gear and paravanes, ignoring any international advances in theory, practice, or equipment. Naval officers assigned to Desk N in the Bureau of Ordnance, which controlled mines, followed suit, storing the Mk 6 antenna mine inventory for future use.[2]

U.S. Navy fleet strategists assigned to develop plans for war against Japan, identified as the next probable enemy, were hampered in their planning by lack of minesweeping experience. Capture of islands in the Pacific as forward bases for a U.S. Navy advance would require minesweepers to precede the fleet for amphibious assaults, but it was an impractical and improbable plan given the number of available ships. Coordination of MCM development and application called for leaders and planners from a navy that could not even field a large enough cadre of mine experts to staff a school of instructors.[3]

This failure to retain active duty, mine warfare expertise severely hampered the development of operational MCM in the postwar period. In yearly exercises known as Fleet Problems, held between 1923 and 1940, U.S. Navy forces trained for warfare in different geographical locations, against a variety of threats, but mine warfare rarely played a crucial role in these scenarios. With few minesweepers and little gear remaining in the fleet, the Navy depended on outdated paravanes for ship protection and assigned aging destroyers to practice only the most minimal precursor sweeps. In almost every case when mine warfare was included in fleet problem exercises, both the mines and the countermeasures were simulated.

Such simulation reaffirmed the image of MCM as a problem easily solved. In 1924, in Fleet Problem 3, Army and Navy defenders threw a curve ball at the attacking forces during the planning process by "constructively" pretending to mine the approaches to the Panama Canal. After considerable difficulty the attacking naval force found two destroyer minesweepers to clear the purported mines but could not locate any sweep gear. After several unsuccessful attempts to devise high-speed sweep gear, the commanding officer of destroyer *Bridgeport* copied the old Russian gear used at Port Arthur in 1904 for a two-ship sweep at slow speed. Battleship paravanes and this preliminary minesweeping revealed the absence of real mines early in the exercise; the attacking commander quickly abandoned all MCM attempts, noting that "in view of the limited time remaining to us, we could afford to ignore the danger of mines."[4]

Although such attitudes "damning" the real effects of mine warfare remained widespread among naval officers, those in high command did not completely ignore the demonstrated need for MCM capabilities in the fleet.

Minesweepers were subsequently added to the fleet composition of later exercises, although most often as tenders and convoy escorts. Simulated mining of Caribbean ports in 1927 exercises promoted the brief return of "Bird"-class minesweepers to MCM duties, but mechanical malfunctions and the regular parting of their sweep gear during the one hour allotted for theoretical clearance did not recommend their continued use. The increased speed of modern vessels, the lack of minesweeping gear for faster destroyers, and the failing condition of the existing minesweepers led to continued dependence on paravanes and avoidance of simulated minefields in many future exercises.[5]

Such inefficient measures caused some consternation to those in high command. A practice landing at Midway in September 1935 revealed "the grave weakness which exists in minesweeping material and methods in the fleet at the present time." This was ascribed to conflicting priorities. "Owing to the heavy demand on the Base Force for services," wrote Commander in Chief, U.S. Fleet (COMINCH), Admiral Joseph Mason Reeves to the CNO, "minesweeping has been largely neglected and until those deficiencies are remedied will in all probability remain neglected to a considerable extent."[6] As a result of this inadequacy in MCM capabilities, exercises after 1935 often included destroyer divisions tasked as high-speed minesweepers. They were not, however, allotted practice mines or sweep gear.[7]

Although mines were included in most later exercises, general ignorance of the subject severely hampered the development of scenarios relevant to the Navy's mine warfare capabilities. Fleet strategists, unaware of advances in mine development and clearance methods during World War I, particularly those refined in the North Sea Mine Barrage, consistently failed to apply any of the lessons of past experience. Ignorant of the nature of American antenna mines, which were the bulk of U.S. mine stockpiles, planners wrongly assumed that paravanes would be effective against all anchored mines. Destroyers assigned to mine clearance were assumed for the sake of the exercise to clear all hazards with a single pass. Furthermore, mining plans were consistently based on the incorrect assumption that all anchored mines were only effective in water less than 100 fathoms deep.[8]

In 1937 the Commander Scouting Force called attention to such misinformation in planning and conducting the mine warfare aspects of these exercises. "In view of the general lack of information in the fleet concerning mines and mining," he reported, "it does seem desirable that the attention of the fleet be called to the fact that the limitations imposed for the Problem are not the actual limitations of our own mines, or those of our probable opponents."[9] This view was echoed by the Commander Minecraft, Battle Force, who in 1940 recommended reconsideration of the effectiveness of paravanes, the limited sweep time and equipment allotted minesweepers, and the need for faster, shallow-draft minesweepers.[10]

As a result of these postwar operational exercises, few naval officers learned much about the real capabilities of modern mines or were adequately prepared to clear them. Fleets involved in a mine warfare scenario, pressed to meet all the conditions of the exercise with inadequate resources, often formulated mining and MCM plans inconsistent with the actual warfare conditions of even a previous generation. Without pressure from the highest levels to study and advance knowledge in all areas of mine warfare, the Navy failed to remember the MCM lessons of its own recent history. " The best we could do," noted one officer faced with the problem in the spring of 1940, "was to recognize probability of mines and pretend that we would sweep."[11]

Within the Navy's scientific community postwar MCM also faced the challenge of competing resources. What little American MCM research continued between the wars centered in the Bureau of Ordnance's small "mine building" test station at the Washington Navy Yard. Established in late 1918 to test both mines and possible countermeasures, this laboratory was converted and reclassified by the bureau in 1929 into an independent research agency, the Naval Ordnance Laboratory (NOL), responsible for a wide variety of ordnance tasks.

Between 1918 and 1939, lean years for the Navy as a whole, Bureau of Ordnance funding for NOL regularly fell short of its needs. Pressed by competing research requirements under postwar reorganization, the bureau transferred funding away from mine warfare research. All mine and MCM research and development thereafter was limited to 20 percent of NOL's mission, with MCM work the smallest portion of that effort.

Despite these limitations, NOL scientists did study a variety of influences that could be used to activate mines, along with possible countermeasures to them. Such influence mines would require no actual contact with the ship's hull to detonate. In addition to magnetic influences, NOL scientists isolated acoustic influences (sound waves made by ships moving in water) and pressure influences (the change in pressure on the bottom made by ship displacement), as well as electrical, optic, seismic, gravitational, chemical, and cosmic ray means of detonating mines. NOL's attempt to actually develop a working magnetic mine, however, met with little success. At its lowest point in peacetime, the entire laboratory staff was reduced to two scientists responsible not only for mine and MCM work but for all U.S. Navy ordnance as well. As a result, MCM suffered from the lack of laboratory facilities dedicated to the specific study of countering mines. As one minor aspect of the mission of an overburdened peacetime laboratory, MCM progress remained stagnant.[12]

In contrast, the British actively pursued mine warfare and MCM in the years after World War I. British MCM specialists made advances in both active minesweeping and passive ship protection techniques. Fine-tuning the *Oropesa*, or "O" sweep, to provide single-ship clearance and developing better gear and offensive tactics for high-speed destroyer minesweeping, British MCM specialists enhanced the small force of commissioned minesweepers the Royal Navy had retained in service.

Building on lessons learned from their own wartime developments and those of the United States in antenna mines, British mine warfare specialists refined magnetic influence mines and magnetic countermeasures. Between 1918 and 1937 they scientifically measured the natural magnetic characteristics, or signature, of ships and developed working influence mines. With their increased sensitivity, discrimination, explosive force, and power projection, early influence mines had the potential to cause greater damage to existing capital ships by breaking their keels and unseating engines. Most influence mines could also be bottom laid, leaving no mooring cables to catch, and thus rendering paravanes and other mechanical minesweeping gear useless.

British scientists centered their magnetic MCM efforts on developing a passive means to protect ships against mines by altering their magnetic signatures. Drawing on earlier studies of magnetic density and the scientific work of mathematician Karl Friedrich Gauss, they soon developed encircling "degaussing girdles" for ships. These electromagnetic coils could be set either to increase a ship's magnetic field, thus exploding mines harmlessly at a distance, or to neutralize the ship's magnetic field, thereby allowing it to pass over the mine. Degaussing altered the ship's induced magnetic signature; "deperming" masked the permanent magnetism acquired by a ship in construction. "Flashing" and "wiping" were similar, although temporary, procedures designed to protect ships from magnetic influence mines.[13]

Disarmed by the Treaty of Versailles at the conclusion of World War 1, Germany secretly began improving its mine warfare capability with Soviet assistance during the interwar years.[14] When war broke out in Europe in 1939, Germany quickly mined the coast of England, as it had in 1914, using its existing stock of contact mines. Putting their paravanes and stern sweeps back in operation and recalling to active duty their well-developed MCM naval reserve force, the Royal Navy was as ready to counter contact mines as any navy could reasonably be.

British minesweeping squadrons began sweeping the coast, but in September 1939 a ship was lost inexplicably in a swept channel. Ship losses increased, and the British quickly added magnets to their sweep cables, correctly suspecting that German aircraft were planting magnetic influence mines. After unsuccessfully attempting to countermine the magnetic mines, the British discovered an intact, undetonated German magnetic mine dropped in error on a beach. Disassembling the mine, experts from HMS *Vernon*, the Royal Navy's mine school, began experimenting with active and passive measures specifically tailored to counter the mine's settings.[15]

British attempts to counter magnetic mines in the months after their discovery took several different forms. Fitting large ships with huge electromagnets, they increased the ships' influence field, generating a false magnetic signature to fool the mine into detonating before a ship entered lethal range. Shallow-water minesweeping by wooden vessels towing magnetized iron bars and by degaussed ships towing magnetized sleds also brought some success. Low-flying passes by Royal Air Force bombers equipped with huge, pulsating magnetized coils proved creative but ineffective. Seeking a more viable method, the British finally developed the towed "LL" sweep—two buoyant electromagnetic cables with both long- and short-tailed electrodes that were streamed by degaussed ships to allow detonation of magnetic mines over a wider path.[16]

Responding to British minesweeping success, the Germans progressively increased the sophistication of their mines. First, they added ship counters, allowing between one and sixteen ships to pass before detonating, making minesweeping more tedious and costly by requiring repeated sweeps of the field. Later in 1939 they introduced acoustically-actuated mines and magnetic-acoustic combination mines. Unlike magnetic fields, the complex noises generated by ships passing through water cannot be muffled easily. To prevent the increasingly sensitive German microphones from detecting ship noise, the British developed noisemakers and, later, oscillators that could be towed by most vessels causing the mines to explode prematurely. German magnetic-acoustic mines, which would detonate only when acoustic and magnetic signatures were detected at the same moment, required simultaneous influence sweeping by both methods.[17]

German use of influence mines required the British to develop new minesweeping tactics. The standard sweep formation for the contact mines of World War I, designated the "A" sweep, married two ships together with the sweep cable between them; but under air attack, a distinctive feature of

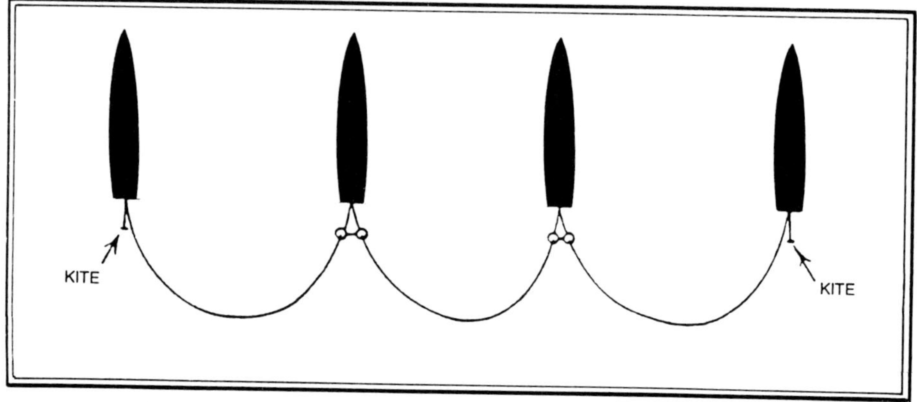

An example of the "A" formation used during World War I. From *Discovery* (January 1946).

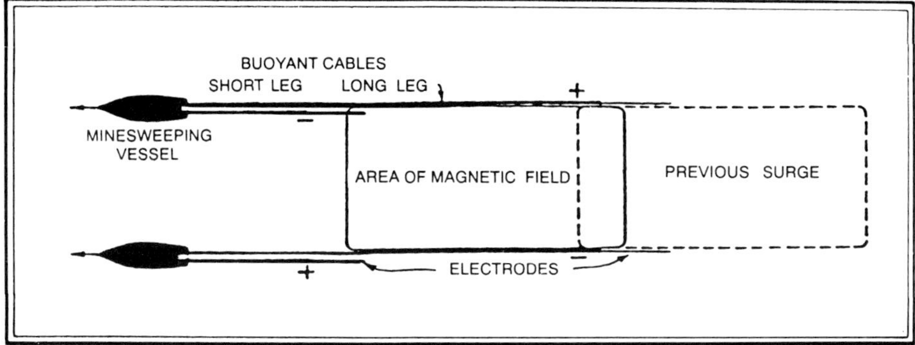

The British "LL" sweep. From *Discovery* (January 1946).

World War II, paired ships could not easily maneuver. Minesweeping in home waters had traditionally been a daytime operation, but air raids soon caused minesweeping forces to shift to night sweeps, making destruction of swept mines difficult. British improvements to the *Oropesa* sweep gear allowed vessels to sweep contact minefields singly, protecting each other in echelon. Under this method the "G" formation became the most widely used, with minesweeping paths overlapping and the lead vessel, when possible, in waters already swept. In assault sweeps only the first ship in line was unprotected. By adding magnetic minesweeping gear and towed acoustic actuating devices to mechanical minesweepers, ships could in theory sweep singly for both contact and influence mines. In actual practice the minesweepers usually found that by steaming in pairs or small groups over the minefield they could intensify the influence fields and clear a larger area more quickly. Influence mines also made intelligence on mine types and placement crucial for the mines to be swept properly.[18]

The United States assisted Britain by transferring old ships and producing new ships to meet specific British needs. Among the minesweeping vessels in the early Lend-Lease program was a dual-purpose, 180-foot escort patrol craft (PCE) designed for antisubmarine warfare and minesweeping. Learning from British experience with influence mines, the United States altered new fleet minesweeper designs in production by 1941, increasing the electrical power necessary for magnetic and acoustic sweep devices.[19]

Between 1939 and 1941, urged on by a few individuals in key positions, the Navy Department restructured its mine warfare program. As CNO, Admiral Harold R. Stark encouraged the MCM interest of Captain Alexander Sharp, a former minelayer commander who led the planning desk of the Naval Districts Division. With authority over mines and countermeasures divided between the bureaus, the Naval Districts Division retained control only over minesweeping for base defense, but by default became responsible for the CNO's entire mine warfare program.

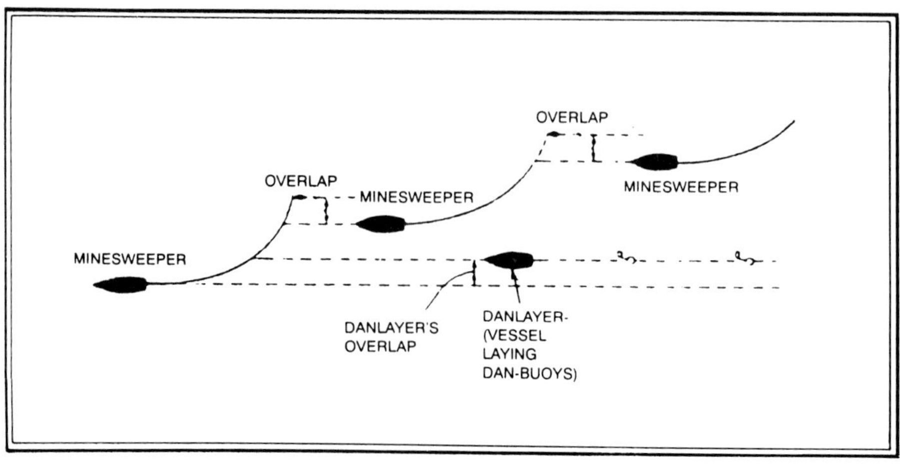

The British "G" sweep. From *Discovery* (January 1946).

Sharp spent much of early 1940 scouting commercial vessels and fishing boats suitable for conversion to base and auxiliary coastal minesweepers (AMC). After selecting and training the first ship and crew, Sharp sent them from port to port to train reserve crews in basic minesweeping techniques. In addition to obtaining ships, Sharp worked with Lieutenant Commander Edward C. Craig, who manned the minesweeping desk in the Bureau of Construction and Repair, to survey available minesweeping gear. They found British influence minesweeping gear to be the most efficient. Actual selection of the ships to serve as sweeps, the minesweeping gear, and the method of degaussing was ultimately a function of the new Bureau of Ships. This bureau, created in June 1940 by combining the former Bureau of Engineering and Bureau of Construction and Repair, assigned Captain Schuyler N. Pyne to this duty.[20]

The Navy also reorganized its research and development policy and laboratory administration. Redesignating its peacetime Desk N into a smaller Section N (Mines, Nets and Depth Charges), the Bureau of Ordnance recalled Commander Simon P. Fullinwider for the second time since retirement to head the new Section N, a subsection of a larger functional research and development division. During Fullinwider's tenure the bureau led a small resurgence of interest in mine warfare by adding the subject to its curriculum at the Ordnance Postgraduate School. In addition to controlling degaussing at NOL, Fullinwider supervised MCM training at the new Naval Mine Depots at Yorktown, Virginia, and New London, Connecticut. The mine section competed for scarce resources with several other newly formed organizations both within and outside NOL, increasing the demands on the section's small staff and funding.[21]

NOL's tiny mine warfare mission expanded quickly after the discovery of German magnetic mines. In response to German mining with these advanced weapons, Stark ordered degaussing of all U.S. Navy ships on a priority basis. By spring of 1940 British MCM advisors had delivered to NOL scientific information on their degaussing experiments plus a captured German magnetic mine. Learning the characteristics and sensitivity of the German mines, scientists in the laboratory centered their MCM efforts on adapting British passive degaussing methods to protect U.S. Navy ships. Recognizing that minesweepers by the nature of their work required more thorough protection from mines than other types of vessels, NOL scientists began developing increasingly powerful degaussing coils and improved ranges to measure the ships' magnetic fields. Their study reinvigorated the laboratory, raising the personnel level from twenty scientists in 1940 to an average of eight hundred in wartime.[22]

The introduction of magnetic mines and passive countermeasures highlighted the need for active coordination and illustrated the kind of turf battles that were features of the Navy's decentralized organization. Between 1939 and 1941 the Bureau of Ordnance, the Bureau of Engineering, and the Bureau of Construction and Repair (Bureau of Ships after 1940) fought over control and development of individual components of MCM. Escalating controversy between the bureaus forced the Secretary of the Navy and the CNO to intervene. Because of NOL's activity, control of degaussing design and measurement was given to the Bureau of Ordnance and installation and powering of degaussing coils was given to the Bureau of Ships. In a separate struggle, minesweeping technology, particularly the development of active influence minesweeping, was retained by the Bureau of Ships. Further confusing this divided responsibility was the fact that magnetic minesweeping by aircraft, still under development, had previously been assigned to the Bureau of Aeronautics. Admiral Stark depended on Sharp to coordinate the efforts of all the bureaus. Still, working relationships between the bureaus responsible for MCM remained so troublesome throughout the war, observers noted that the only solution was for personnel to bypass official channels.[23]

The Bureau of Ships set up a research group and established a Mine Testing Station on Solomons Island, Maryland, in Chesapeake Bay to test new minesweeping equipment, develop new sweep instructions, and conduct new MCM research. The Solomons facility consisted of only two test vessels that were used for both mine and torpedo countermeasures experiments; it was not, however, a laboratory or a dedicated MCM testing facility. Between 1939 and 1942 the bureau produced few new countermeasures, preferring instead to test and refine British equipment. Lieutenant Commander Hyman G. Rickover, one of the few naval officers who had previously commanded a minesweeper, brought back a small section of sweep cable from England and adapted the British magnetic sweep cable to American minesweepers at his lab in the Electrical Section of the Bureau of Ships. Although the bureau ordered sufficient gear to outfit available minesweepers, inadequate replacements were procured. A few early losses wiped out all reserve gear until nearly the end of the war.[24]

Another mine warfare priority was the designation and outfitting of minesweeping vessels. The few old steel "Bird"-class minesweepers in commission had obsolete and deteriorating mechanical minesweeping equipment, but by June 1940 the new Bureau of Ships had degaussed them and added magnetic sweep equipment to their mechanical gear. Others of their class were recommissioned. A number of older destroyers and submarine chasers were converted for tactical use as fast assault minesweepers (DMS). Sharp's thirty-five large wooden fishing vessels, mostly tuna boats (also given bird names), worked so well that many of their features were incorporated into the Navy's new fleet of seventy 1941 *Accentor*-class auxiliary coastal minesweepers, 97-foot craft capable of speeds up to ten knots. Smaller fishing vessels became yard patrol craft and harbor minesweepers, and their capabilities influenced the design of a class of small wooden yard motor minesweepers (YMS), 116 of which were laid down before 1941. In addition to the YMSs built for future American needs, many YMSs were constructed for other navies under Lend Lease.

Construction began almost immediately on two experimental models of a 220-foot steel fleet minesweeper prototype for the *Raven* and *Auk* (AM) classes. To crew these new and converted vessels, Chief of the Bureau of Naval Personnel Rear Admiral Chester W. Nimitz agreed with Sharp's request to recall reservists to begin immediate training.[25]

Largely through Sharp's vigorous coordination, a sense of direction coalesced the U.S. Navy MCM program by 1941. After completion of its degaussing studies in 1941, NOL turned most of its assets away from MCM to mine development, and responsibility for degaussing shifted to the Naval Districts Division. Personnel assigned to the growing mine force experimented with new sweep methods and equipment at the Naval Mine Depot at Yorktown, and Sharp soon gained the cooperation of both the Bureau of Ordnance and the Bureau of Ships to set up a separate Mine Warfare School nearby. Students in the first class, which graduated in May 1941, used school ships to practice clearing live mines, the first such exercise in over two decades. These graduates became the first commanding officers of the expanded minesweeping force. Between 1940 and 1945 the school trained over 10,000 naval reserve officers in all aspects of mine warfare, from electricity and electronics to mine operation and countermeasures.[26]

At Sharp's suggestion his successor in OPNAV, Lieutenant Commander R. D. Hughes, was assigned in 1941 to a new mine warfare desk in the Naval Districts Division (OP–30-C) with authority restricted to defensive mine warfare and countermeasures. A graduate both of the Naval Academy and the first Mine Warfare School class and an active student of British mine warfare, Hughes obtained assistants in mine disposal and minesweeping and soon became the Navy's principal mine warfare expert. Reorganization the following year reduced his desk to a Mine Warfare Section in the Base Maintenance Division under a senior officer.

His official title and OPNAV position reflected little of his true importance to the overall MCM effort for, as Sharp had before him, Hughes acted as informal liaison between all units to promote cooperation. Throughout the war he remained the driving force behind the U.S. mine warfare program. His Mine Warfare Section centralized coordination of all the diverse MCM projects throughout the Navy by analyzing operational requirements and providing information and direction to the bureaus on mine warfare needs. The section collected information on mine warfare in each phase so that lessons learned were applied in subsequent operations, and it supervised the training and distribution of mine warfare personnel. From 1941 to 1945 this office also produced a classified professional journal, *Mine Warfare Notes*, to keep all personnel in the different bureaus and sections abreast of new technical developments, intelligence, and operational experiences.[27]

Minesweepers assigned to the Fourteenth Naval District (Hawaii) gave the first notice of the Japanese attack on the U.S. Pacific Fleet at Pearl Harbor when they sighted a periscope during a routine, early morning, minesweeping operation. During the ensuing action the senior mine warfare officer, Commander Mine Division One, lost his ship, *Oglala* (CM–4), while the minesweeper *Tern* (AM–31) rescued sailors and fought fires on battleships *Arizona* (BB–39) and *West Virginia* (BB–48). Although the minesweepers were not the focus of the Japanese attack and suffered relatively little damage, their sailors joined in Pearl Harbor's defense by manning every gun they could find and bringing down some Japanese bombers.[28]

Navy-wide unfamiliarity with MCM tactics and the complexities of integrating the mine force into fleet operations clearly called for creative MCM leadership in wartime, but the U.S. Navy had done little to develop a mine warfare leadership cadre. After Pearl Harbor, Admiral Stark did the next best thing by assigning mine warfare liaison officers to the individual fleet staffs to integrate information on mines and countermeasures into tactical planning and operations.

In early 1942 German U-boats laid over three hundred influence mines in small fields off Delaware Bay; Chesapeake Bay; Jacksonville, Florida; and Charleston, South Carolina; closing the ports for several days and effectively restricting fleet MCM units to eliminating the immediate threat in home waters. In June German submarines remined the entrance to Chesapeake Bay and Hampton Roads, Virginia, sinking two ships, damaging one, and sealing off vital Norfolk naval ship traffic for four days. Despite thorough sweeping by ships from the Yorktown Mine Warfare School, SS *Santore* hit a mine and went down in an area already believed cleared, causing sweepers to tighten up operations to counter German remining. Although the Navy hastily converted 125 fishing trawlers to minesweepers, some American ports were closed by mines for over a month.[29]

Forced to stretch limited assets to provide quick and effective clearance for several important ports, the Navy in 1941 adopted nearly wholesale the British practice of clearing "Q" routes, that is, specified channels between harbors that needed to be constantly surveyed and swept for protection against both submarines and mines. At twenty-eight U.S. ports and harbors, Army, Navy, and Coast Guard forces strung large steel nets across the harbor entrances to deter enemy submarines and rigged large booms to defend against attack on the surface, all MCM measures of an earlier age. To patrol the coastline and to provide adequate home-port MCM capability, the United States ultimately employed over 260 fishing boats converted to minesweepers on combined harbor patrol and minesweeping duty.[30]

Once the Japanese attacked Pearl Harbor, America began to develop a large-scale MCM force. Immediate congressional budget increases and escalating German air and submarine minelaying allowed the Navy to spend more money on MCM, particularly on new ships and gear. New minesweeper construction took advantage of available wood and woodworking capabilities in U.S. yacht and boat yards, rather than of improved theories of MCM ship design, and relied upon the lines of the converted trawlers and minesweepers of World War I. Depending on degaussing for passive protection of all minesweepers from magnetic mines, the Bureau of Ships built both steel minesweepers and wooden vessels constructed with substantial amounts of iron and steel equipment, and did not seek to design ships with nonmagnetic qualities and materials for protection.

The post-Pearl Harbor building program also included the 184-foot *Admirable*-class AM, a steel, 14-knot ship that was mass produced as a minesweeping, patrol escort, and degaussing vessel. These ships were good minesweepers with auxiliary engines powerful enough to generate the electrical power required for magnetic sweeping. The smaller YMSs, wooden-hulled ships originally designed for coastal operations, proved capable of countering magnetic mines and adaptable to the requirements of overseas deployment. The YMSs were so popular with the U.S. and foreign navies that they accounted for almost all influence clearing and exploratory sweeping, yet they remained vulnerable to increasingly sensitive magnetic influence fusing technology. Concern for the damage caused by mines had similarly little effect on wartime capital ship design, and ship protection against mines while underway continued to depend on degaussing and paravanes.[31]

Additional minecraft and expanded missions called for reorganization of

National Archives 19–N–55263

Notable (AM–267), an *Admirable*-class auxiliary minesweeper.

the mine force. The Base Force, U.S. Fleet, was abolished in February 1941 and became Base Force, U.S. Pacific Fleet. There was no actual change, however, in the command structure from the old Minecraft, Battle Force, although additional minecraft were added to the Base Force, U.S. Pacific Fleet, in anticipation of war. Individual minesweepers reported to Training Squadron 6 (later Service Squadron 6). Minelayers were originally assigned to Commander Minecraft, Battle Force, U.S. Pacific Fleet, but this command was abolished in a Pacific Fleet reorganization of April 1942, and the units were combined with the minesweepers in Service Squadron 6.

In August 1942 active and reserve minesweepers in the Atlantic Fleet came under the administrative control of Vice Admiral Sharp as the new Commander Service Force, Atlantic, through Commander Service Squadron 5 at Norfolk. Quickly preparing his fleet for combat operations, Sharp reorganized minesweeper administration, readiness, and training for deployment and generally acted as type commander. Until October 1944, however, there was no specialized mine warfare command or centralized

National Archives 80–G–K–1713

The crew of battleship *New Mexico* (BB–40) deploys a paravane in the Pacific Ocean, 1944.

control of mine warfare vessels in the Pacific comparable to the Atlantic Fleet.[32]

Development of MCM tactical employment continued to suffer from a general lack of line officer knowledge of mine warfare. To promote better interaction between civilian scientists studying MCM techniques and the naval personnel employing practical MCM methods, Dr. Ellis Johnson, a scientist at NOL, formed a wargaming group in early 1942. Scientists from NOL and minesweeping experts from the Bureau of Ships met in the evenings to develop hypothetical offensive and defensive mine situations and to formulate MCM strategy. In 1942 this wargaming group combined with a parallel study group in the Bureau of Ordnance, and moved to OPNAV's Mine Warfare Section as the Mine Warfare Operational Research Group. Most active in 1942 and 1943, the Operational Research Group studied German mining tactics, planned offensive mining strategies, developed effective countermeasures, and solved several mine disposal, sweeping, and degaussing problems.[33]

Operational experience against contact and magnetic mines in support of Allied amphibious operations during the Mediterranean campaign reintegrated the reformed U.S. minesweeping force into the fleet for the first time since World War I. Because of the enemy's preference for contact and simple magnetic mines, these minesweeping operations remained nearly as technically uncomplicated as those of interwar fleet exercises. *Raven* (AM–55) and *Osprey* (AM–56) did exploratory assault sweeping for the invasion of North Africa in November 1942 and later served as landing craft control vessels. Allied landings at Sicily, Salerno, and Anzio pitched U.S. minesweeping forces into assaults against both sea and shallow-water land mines. At Salerno insufficient time allotted to clear the minefields resulted in the loss of one ship. In the landing of U.S. forces on the beaches at Anzio-Nettuno in January 1944, a large group of twenty-three minesweepers swept the approaches in time for the assault force's arrival and found that maneuvering in the crowded waters made ship traffic a greater hazard than the mines themselves. Minesweepers operating in this theater between 1942 and early 1944 were thus able to relearn crucial MCM lessons about the need for prior planning, properly equipping minesweepers, and control over their own forces.[34]

National Archives 19–N–23990

Osprey (AM–56) took part in the preparatory assault sweep for the Allied invasion of North Africa in November 1942.

The greatest amphibious challenge to the Allied forces fighting Germany was the invasion of France. Mines were a significant factor in preinvasion planning. Although constant Allied night attacks in the months preceding the landing at Normandy had prevented intensive German minelaying in the English Channel, mines were expected to pose a bigger problem in French harbors. Inaccurate early intelligence for a planned invasion at Normandy indicated that contact mines, not influence type mines, were the biggest threat. In reality the real challenge to the MCM forces off the coast of France turned out to be a new type of influence mine. These particular pressure mines, called "oysters," were developed in Germany in 1940 and were activated when the pressure on the mine dropped at least two inches for at least seven seconds—the average pressure drop caused by a 120-foot ship passing at ten knots.

Although there was neither a known countermeasure for these mines nor a passive or active measure guaranteed to deceive the mines into prematurely firing, the British believed that they would not detonate when ships passed them slowly at four knots. Fearing that the Allies would exploit, counter, and use pressure mines against the German Navy, the German Naval Staff had withheld them from service for most of the war. In early 1944 Adolph Hitler personally ordered that 4,000 pressure mines be sown around Normandy, Le Havre, and Cherbourg to block Allied invasion attempts.[35]

British information on possible German pressure mines led COMINCH/CNO Admiral Ernest J. King to direct the Bureau of Ships in early 1944 to develop countermeasures against the German mines. Although the threat of pressure mines was known earlier, little progress to counter them had been made. Now, with King's interest, scientists at the Solomons Island test station developed a displacement sweep in only sixteen days. This device, a large towed barge designed to explode pressure mines by creating the change in hydrostatic pressure needed to fire them, could not, however, protect the ship towing the barge.

As scientists worked to correct this problem, the fleet tried alternate approaches to counter pressure minefields. Through statistical analysis the British identified possible safe speeds at which different types of ships could slowly pass over pressure mines without detonating them. Some destroyer commanders planned to fire these mines by steaming toward them at high speed and making a full rudder, high-speed turn away from the field, propagating a swell toward the minefield. This procedure usually proved too hazardous or, as in the case of combination settings, ineffective. Countermining also failed to actuate pressure mines. Until actual German pressure mines could be recovered and exploited, Allied forces had to rely on intelligence estimates of the mines' potential.[36]

Thus the Allied fleet neither expected to meet much opposition from pressure mines nor had developed an effective method to counter them. In advance of the assault on Normandy, 245 Allied minesweeping vessels swept approaches from the English coast toward the landing area with standard contact and influence minesweeping gear, marking paths with both lighted and sonic buoys. Because of concurrent American naval actions in the Pacific, the British provided the preponderance of naval forces for the landing. Of the 306 minesweeping vessels ultimately in the invading force, only 32 were from the U.S. Navy—11 *Raven*-class AMs and 21 YMSs, all in the American sector of the proposed landing area.

Consistent with intelligence estimates, the Germans had indeed mined the channel close to Cherbourg with a line of contact mines, some with antennas. Most of these mines had, however, passed their timed life cycle and were inert by June 1944. There was little hint during the approach that more advanced influence mines awaited the invading force in harbor waters.

As the American forces swept the English Channel, moving toward Normandy, rough weather made complete clearance impossible. On 5 June *Osprey* hit a mine and sank while sweeping with U.S. Mine Squadron 7. The combined U.S., British, and Canadian mine force approaching the American sector had difficulty navigating accurately in the rough seas and darkness. The minesweepers preceding the landing force swept tightly in formation. On approaching Utah Beach on 6 June 1944, the invasion force detonated magnetic mines, losing several combatant ships and at least sixteen landing craft.[37]

Despite mechanical and influence minesweeping, Allied losses to mines at Normandy at first seemed a balanced risk. Only after the landing did minesweeping and minehunting reveal massive minefields containing

National Archives 80–G–651677

Tide (AM–125) sinks after hitting a mine off Omaha Beach, 7 June 1944, one day after the Normandy invasion.

hundreds of pressure and combination pressure-acoustic mines that the landing force had luckily missed.

Throughout the following summer, Allied minesweepers worked to develop a method to defeat the pressure mines. Observing that summer storms often created sufficient pressure swells to set them off, they began sweeping acoustically for combination pressure-acoustic mines during such weather. To defeat the weather, the Germans, whose aircraft continued mining the Allied-controlled waters, merely readjusted their pressure settings for coarser sensitivities.[38]

Analysis of German advanced combination mines found in Cherbourg, captured by the U.S. Army on 27 June, made the full extent of the German mine defenses clearer and more daunting and changed the focus of Allied MCM operations and policy. The mixed minefields found at Cherbourg contained thousands of contact, magnetic, and pressure mines, some fitted with ship counters. Intermingled with shipwrecks and underwater obstructions were also unused ordnance and clever "Katie" mines, which were concrete-encased charges on tripods rigged so that sweep wires would simply roll the tripods aside, allowing the mines to right themselves to hit a real ship. Clearance required divers, salvage efforts, and constant, repetitive sweeps for all types of mines. Allied forces off Cherbourg swept eight times a day for eighty-five days and found that the majority of the thickly laid mines in and around the harbor required multiple actuations or planting of individual countermines by divers.[39]

Although largely missed at the time, the meaning of the experience at Normandy for MCM was profound. Influence mines had radically altered the MCM requirements of most navies. No longer could minesweepers proceed into unknown waters towing only one type of sweep gear to meet the threat of a simple mine. The only true counter to increasingly complex mines and mixed minefields was minehunting: the identification of minefields and the exact characteristics of the mines within them, a costly and difficult process. Active hunting of such minefields by any MCM force required close integration of emerging technology.

The sharp increase in the numbers of pressure and combination mines found at Normandy and Cherbourg did, however, immediately challenge the MCM priorities of the Allied forces. As American ships joined in the long-term effort to clear the mines from French waters, Admiral King ordered that advanced MCM development be made the highest priority and that all laboratories and bureaus develop research and construction projects to address the problem posed by sophisticated mines. Hughes and the OPNAV Mine Warfare Section continued their efforts to coordinate the different programs within the U.S. Navy.[40]

In this redoubled effort American and British scientists centered their work on traditional sweeping of these advanced mines with combinations of actuating devices that used water-filled balloons and displacement barges to trigger the mines' pressure mechanisms. At NOL tugs towed the "Loch Ness Monster," a huge nylon sleeve with a very large mouth forward and a small opening aft that would take in a large volume of water which accelerated within the sleeve and departed the after end fast enough to reduce pressure in its wake. The device was later sent to the Pacific Fleet to sweep American pressure mines but was never operationally tested. Good in theory, the Loch Ness Monster proved physically unmanageable.

Scientists at the David Taylor Model Basin developed a towed barge called an "egg crate," and those at the Bureau of Yards and Docks developed a similar "cube steak." Unfortunately these towed barges cleared few mines when tested in European waters. Other scientists tried to alter water pressure with countermines.

The only successful and practical scheme developed to counter pressure mines by running something ship-like over them was the use of expendable "guinea pig" ships—Liberty ships padded and filled with buoyant materials and remotely piloted from above decks or from the deck of another vessel. The Germans used such ships throughout the war to clear paths through influence minefields, whereas the Allies used them mainly to check cleared passages.[41]

Advances in sonar and magnetic detectors showed promise for minehunting application. Scientists at the Navy Electronics Laboratory in San Diego, California, installed active sonar mine detection units aboard some U.S. submarines, allowing them to prowl safely through Japanese fields of moored

Courtesy U.S. Naval Institute

The skeleton crew of this "guinea pig" ship uses helmets and mattresses above decks for shock protection from influence mine detonation.

contact mines. The Navy also experimented with towed underwater magnetic mine detectors, and mines found were usually disposed of by trained explosive ordnance disposal (EOD) divers. EOD units and underwater demolition teams (UDT) were also deployed to reconnoiter beach approaches for mines on assault landings. U.S. Navy divers and Navy hydrographic units, who often accompanied minesweeping task groups in the Pacific to chart and mark swept channels, were often integrated into MCM operations.[42]

Minesweeping vessels continued to lead advancing U.S. naval forces in all theaters during the last year of the war. In the Mediterranean in 1944, most minesweepers swept for contact mines, but the real advance in Mediterranean MCM tactics was the use of aircraft to vector ships through minefields in shallower waters.

Lighter-than-air vehicles, particularly blimp *K–109*, guided the U.S. Navy's Mine Squadron 11 and Mine Division 18 through the Mediterranean minefields, ultimately assisting in the identification of eighty-six mines in these waters. Blimps were the first U.S. air assets to spot mines successfully for minesweeping forces. At Key West, Florida, in November 1944 a Navy blimp also participated in mine destruction. Over a seven-day period, one of these airships used a .50-caliber machine gun to sink twenty-two mines raised to the surface by minesweepers in the first operational combined air and surface mine clearance operation in the U.S. Navy.[43]

Until the U.S. Navy began offensive operations against the Japanese in the Pacific, mine warfare on both sides in that theater proceeded haphazardly. The Japanese had both the ability and expertise to sow deep sea fields with improved contact mines set at depths of 1,500 to 3,500 feet, yet they rarely did, preferring to use the weapons defensively to protect their own coasts. Using a mixture of German-type magnetic mines and old British contact mines from World War I, the Japanese apparently either plotted their minefields poorly or navigated their ships badly; their mines sank four of their own vessels in Pacific waters before 1942.[44]

U.S. submarines began mining the Pacific in 1942 with both moored contact and early magnetic mines. U.S. magnetic mines planted off Bangkok, Haiphong, and the Hainan Strait in particular were extremely effective, claiming one victim for every eight mines laid. Of six U.S. mines laid at Haiphong by British forces in 1943, three sunk ships and the remainder kept the Japanese out of that port throughout the rest of the war.[45]

As Navy planners developed their Pacific strategy to defeat the Japanese Navy, they expected minesweeping vessels to lead amphibious assault operations, and the various laboratories and MCM units responded with improved equipment. Pacific Fleet minesweeping task groups consisted of new YMSs, new and converted AMs, and DMSs designed to conduct exploratory moored sweeps at high speeds in assault forces. To allow these ships to clear shallow waters up to the shoreline, scientists at the Solomons test station used captured German fast minesweeping equipment to develop lighter "5–G" sweep gear. NOL's experiments with a minehunting underwater ordnance locator nicknamed "King Kong," a 6-foot towed electromagnetic detector used to find lost ordnance buried in the silt off the Yorktown test station, led to the development of magnetic devices capable of detecting everything from mines and small arms to submarines. Navy-wide improvements in navigation, particularly in the development of the dead reckoning tracer and navigational radar, assisted forces in the Pacific by allowing MCM vessels to maintain a safer, steadier track through minefields and to mark swept paths precisely.[46]

Concerned by rumors of advanced Japanese mine technology, Admiral King continued to press for high priority development of maximum MCM capabilities for the advancing Pacific Fleet. Stung by reports prepared for him by British experts imported to analyze American MCM, which cited the U.S. Navy's "lack of preparation in the field of mine location," King personally selected the best personnel to man the Navy's new mine disposal units. To assist ships and divers in locating advanced mines in the Pacific, King directed the Bureau of Ships to develop a program based on the Underwater Ordnance Locator produced by General Electric and ordered these locators placed on Navy ships.[47]

Pacific Fleet minecraft, both minelayers and minesweepers, had remained under the command of Service Squadron 6 since 1942. That organization had grown immensely in size and, as one participant noted, there was no "centralized control" of mine warfare vessels in that fleet. In October 1944 Vice Admiral Sharp voluntarily gave up his third star and reverted to rear admiral for the opportunity to command the first MCM type command, Minecraft, Pacific Fleet. With that command came the unprecedented opportunity for a dual administrative and operational command, as Sharp also became the operational commander of Task Group 52.2, Mine Flotilla, for the advance on Japan. With Sharp's minesweepers in the vanguard, U.S. forces reentered the Philippines, sweeping hundreds of contact mines while assault vessels cut many more with their paravanes.[48]

Sharp's forces expanded as he moved quickly through the Pacific, participating in most major operations along the way, including Leyte, Lingayen Gulf, Corregidor, and Iwo Jima. By the time he reached Okinawa, Sharp commanded the largest minesweeping fleet ever assembled by the U.S. Navy. Seventy-five minesweeping ships and 45 assisting ships scoured over 3,000 square miles off southern Okinawa for a week in advance of the 1 April 1945 landing, accounting for 222 contact mines. Sweep operations continued during the two-month contest despite repeated kamikaze attacks. Four of Sharp's minesweepers sank at Okinawa and 16 were damaged, but his MCM force accounted for over 510 contact mines swept and 95 aircraft shot down.[49]

Minesweepers in the Pacific faced their biggest test in the last offensive

National Archives 80G–303838

Motor minesweeper *YMS–362* off Iwo Jima during that invasion, February 1945.

minesweeping operation of the war. Until mid-1945, the only significant obstacles in the sweepers' path had been contact mines laid by the Japanese. While preparing for the amphibious assault at Balikpapan, Dutch Borneo, in June 1945, Pacific Fleet minesweepers for the first time faced a large body of influence mines laid by other U.S. and Allied forces. Surprisingly, this was the first time that American MCM forces encountered American influence mines, and the result shocked the mine force. In a crucial preassault sweep covering sixteen days, the ships, mostly YMSs, swept four successive segments of the minefield, clearing Japanese contact and British and American influence mines planted in increasingly sensitive waves as they neared the shore. The little YMSs, although made of wood, housed magnetic engines and equipment, and their signatures made them dangerously vulnerable to the more advanced Allied mines. To the horror of the minesweeping fleet, seven of these capable ships were sunk. Overall, the sweep of Balikpapan was a resounding success for the amphibious forces, which arrived to find a path clear for the assault. The MCM forces recognized, however, that the high cost in minesweeper casualties was a clear sign that they were ill-equipped to deal with advanced mines, even those of their own design.[50]

As U.S. naval forces closed in on Japan in 1945, they began to plan for another large-scale offensive mining campaign to end this war as they had the last one. Operation Starvation, the strategic aerial mining of Japanese coastal and inland waters, concentrated on closing the Shimonoseki Straits to seal off Japan from Asian food supplies and major shipping routes. Immediate plans for minelaying were complicated by the need to determine the type of mines to be used, planting methods, and the countermeasures required to clear them. To be effective, the mining of Japan had to be fast, efficient, and difficult to counter. Army Air Force bombers delivering the most advanced American combination influence mines were able to lay the fields quickly and effectively. According to the Hague Convention of 1907, however, the United States could not legally plant mines that their forces could not sweep.[51]

With the effort to develop countermeasures already pressed to the limit, Fleet Admiral King sought a solution to comply with the Hague Convention. Ultimately, King authorized the use of some virtually unsweepable American mines, including combination mines and acoustic mines sensitive only to low frequency vibrations, but ordered that they be equipped with timed sterilizers designed to turn the mines off or to render them inert. U.S. Army B–29s planted 12,135 American influence mines in Japanese waters in successive waves. Another 13,000 mines were laid by aircraft and submarines over a wide area of the Pacific. Borrowing yet another British practice, the Bureau of Ordnance sent a Mine Modification Unit of experts to alter mine settings on station, making the mines increasingly sensitive in order to confuse Japanese clearance teams. As Japanese minesweepers swept, the bombers returned and remined swept waters with different and more deadly mines.[52]

Japanese attempts to counter the American mines using traditional MCM means had only limited success. They tried mine spotters, magnetic and acoustic sweeps, countermines, radar, searchlights, sweep nets, and divers and apparently also had some ships run through fields with only a bow watch armed with rifles. Continuous remining, aided by the tracking of Japanese countermeasures efforts by U.S. aerial photography, sank or damaged 670 Japanese ships, accounting for a significant portion of the maritime trade. After the war knowledgeable Japanese naval officers conceded that the mining of Japan and the significant reduction of merchant vessels by losses to mines had helped strangle the nation.[53]

By the time the United States dropped the atomic bombs ending the war, the American offensive mining operation in Japanese waters had succeeded in proving the effectiveness of mine warfare. In the aftermath of victory, however, the deeper lessons of the U.S. Navy's mine warfare operations in the Pacific were, to one extent or another, quickly obscured by this very success. The massive minelaying effort had redefined mines as the weapons of powerful nations, but few outside the MCM community would remember the much more extensive effort required to clear the advanced influence mines only recently deployed.[54]

The need for effective mine countermeasures did not end with the surrender of Germany and Japan. The extent of mine clearance efforts by each nation after the war depended in large part on national priorities. Allied forces worked assiduously to clear the heavily mined waters of Northern Europe and the Mediterranean, learning much in the process about the nature of various types of influence mines discovered. U.S. Pacific Fleet MCM ships, forced to clear paths through thousands of Japanese contact and American influence mines to retrieve Allied prisoners of war and to land occupying forces, found the task far more difficult than had been anticipated by those who estimated that mine clearance would take one year to complete.[55]

Throughout the war the U.S. Navy had prepared to meet the threat of advanced influence mines by seeking to develop effective countermeasures for each type and setting. Operational forces, however, only rarely encountered them, and by October 1945 U.S. minesweepers had acquired only limited experience in clearing influence mines, having swept over 10,173 contact mines, but only 467 influence mines, in all theaters.

Clearance of over 25,000 influence mines remaining in Pacific waters was a formidable task facing American minesweepers. Lacking the forces and technology to stage a massive sweep of all the mines, the United States formed a policy of sweeping only essential waters "on a minimum risk basis for U.S. forces." To counter mines lacking sterilizing features, Sharp's successor, Rear Admiral Arthur D. Struble, attempted aerial countermining without notable success, and finally resorted to guinea pig ships to accomplish their limited mission.[56]

American naval officers in command of the sweeping operation greatly underestimated the time required to sweep mines of such complexity in Pacific waters by traditional sweep methods. To complete clearance, the existing Japanese minesweeping forces were ordered to sweep waters not deemed essential to U.S. forces. Beginning in September 1945, 350 small Japanese vessels, under the command of Captain Kyuzo Tamura and assisted by U.S. vessels, swept constantly in the coastal waters around Japan. All U.S. assets were withdrawn in May 1946 after reportedly sweeping 12,000 mines in 22,000 square miles. One Navy official estimated in 1946 that it would take about two years to complete the clearance of mines in Japanese waters. Thereafter the small Japanese minesweeping fleet continued methodically sweeping under the authority of Commander U.S. Naval Forces, Japan (later Commander Naval Forces, Far East [COMNAVFE]). In 1971, estimates noted that after twenty-six years of sweeping, more than 2,000 sensitive influence mines remained in shallow waters.[57]

Few Navy leaders realized how much of their success in countering mines during World War II was due to adaptation of British MCM technology, tactics, and applications. The Royal Navy by contrast did not forget that in both world wars mines had damaged more British ships than torpedoes. Even as they demobilized their 700-ship coastal minesweeping force, the British continued their MCM shipbuilding program, laying up partially completed new minesweepers and minehunters for future employment.

In the inevitable postwar demobilization the U.S. Navy faced similarly tough questions concerning ship retention and future missions. During the war over 950 ships, including 283 fleet minesweepers, joined the MCM fleet.

Only 58 were casualties; the relatively low number for a dangerous occupation reflects the low number of influence mines faced. But major questions of service unification, establishment of the Navy as the forward line of defense against advances by the Soviets (quickly identified as the next major military threat), and incorporation of nuclear weapons overshadowed mine warfare matters and prevented them from featuring prominently in postwar restructuring.[58]

Planning for offensive warfare against Soviet submarines revolved around air, surface, and undersea assets; it put high priority on the development of hunter/killer ships, submarines, and antisubmarine warfare forces. In attempting to reduce the Navy to a balanced mix of forces that could adequately counter the expected threats, Navy planners included minesweepers as one element of the balanced force needed for amphibious operations. Basing their planning on the overwhelming success of minesweeping forces in World War II operations, and not on lessons learned by the mine force in operations such as those at Balikpapan, planners failed to note that even when using all known conventional MCM methods, U.S. forces had more trouble sweeping their own mines laid in Asian waters than those of the enemy.[59]

Other lessons went unlearned because few experienced men remained in the mine force to remember them. Few naval officers had experiences comparable to those of Alexander Sharp, who had personally reordered wartime MCM priorities in OPNAV, then successively served as type commander for MCM ships in both the Atlantic and Pacific Fleets, and did double duty as operational commander of minecraft in the Pacific. When the war was over, Sharp retired, and the core of the MCM force, the reservists, returned to civilian life. Once again mine warfare became the province of the few interested individuals remaining on active duty.

In the postwar reorganization of the fleet seven type commands were created, each responsible for the condition and readiness of different types of ships. This restructuring increased Navy professionalism through subspecialties within many individual communities Navy-wide, including MCM. In the Atlantic Fleet in April 1946 a type command for minecraft, known as Mine Force, Atlantic Fleet (MINELANT), replaced the MCM portion of the Service Force, creating a parallel organization to Minecraft, Pacific Fleet. Yet a few months later in January 1947, in further reorganization, CNO Fleet Admiral Chester Nimitz reduced the Pacific Fleet MCM forces to a handful of ships and eliminated the Minecraft, Pacific Fleet, type command. Responsibility for minesweeping in the Pacific then became a collateral duty of the service and cruiser-destroyer forces.[60]

The wartime U.S. Navy MCM force successfully countered mines and materially assisted in crucial amphibious assault landings with relatively little loss. Except at Balikpapan, where advanced influence mines took the heaviest toll on the minesweeping fleet, such success seemed on reflection easily gained. But underpinning that success was an element of luck; had more advanced influence mines been present in the American sweep sectors at Normandy and Cherbourg or had the Japanese laid more of their own influence mines throughout the Pacific, the lessons learned by the Navy as a whole might have been far different from the ready mastery over mines remembered immediately after 1945.

More thoughtful observers reviewing the Navy's mine warfare program in 1946 pointedly asserted that solving the MCM problems that remained after the war would require additional study, resources, and staff. Recommending that a high priority peacetime MCM program be developed to meet the challenge of countering advanced influence mines, these observers included prevention of minelaying, development of mine locating devices for minehunting, and new minesweeping equipment among items of immediate concern. Noting the absence of active duty naval officers from mine warfare, these observers also suggested establishment of a high-visibility program to familiarize all military officers with mine warfare matters through better officer training programs at staff colleges, service academies, and flight schools.[61]

Throughout the war the Navy succeeded in developing a quick-fix approach to MCM that allowed another generation of U.S. naval officers to "damn" the effects of sea mines. But in a transition unnoticed by many of these officers, advanced influence mines, particularly pressure mines that required accurate intelligence on mine type and placement for effective clearance, changed the focus of MCM from a low-profile profession with easily fabricated solutions to a complex one that required a long-term focus. Influence mines had immeasurably magnified the potential of mine employment against the U.S. Navy. Yet this critical lesson, made so clear by wartime experience, went largely ignored at war's end, with inevitable consequences for the combat ahead.

3

The Wonsan Generation:
Lessons Relearned
1945–1965

Americans rejoiced in success after World War II, celebrating the end of war and the return to a peaceful world. With victory came interest in evaluating the success of recent operations, but the focus of American attention on mine warfare in the postwar period remained steadily on the contributions of effective minelaying campaigns and the proficiency of the minesweeping fleet in the Pacific. Only the mine force remembered the devastation caused by influence mines at places like Balikpapan. American inability to clear advanced influence mines even of its own design at Balikpapan went largely unnoticed in the celebration of victory.

The successful operational MCM efforts in World War II limited MCM development after the war. Despite rapid advances in influence mine development, the largest mine threat throughout the war remained the antique, World War I moored contact mine. To a budget-conscious Navy, wartime dependence on this cheap, easily countered mine by foreign navies made continued funding for influence MCM all the more unnecessary. On the other hand, the officers assigned to mine warfare in the postwar Navy were well aware of the limits of ships built and equipped to deal with the threats posed by increasingly sensitive magnetic mines in World War II. They recommended the design of improved, MCM-specific, nonmagnetic ship types, additional training, improved tactics, and new research to counter the latest influence mines in development.

However, proponents of MCM readiness had trouble making themselves heard in peacetime. Naval reservists had made up 90 percent of the wartime MCM force and their return to civilian life stripped the MCM population. With a reserve program to establish and other training priorities facing it, the Navy placed little emphasis on educating, promoting, or enhancing the status of the MCM personnel who remained on active duty in the postwar period. In the 1945 reorganization of OPNAV, the Mine Warfare Section, which had been under the new Deputy Chief of Naval Operations (DCNO) for Logistics, moved to the DCNO for Operations, where it competed with ranking proactive elements of the destroyer and submarine communities for attention and appropriations. With peace, mine warfare once again became an unattractive specialization within the Navy.[1]

As a reflection of the mine warfare community's declining influence, experimental MCM advances in progress at war's end were shelved in peacetime. The first successful minehunting sonar, designated QLA and installed on nine U.S. submarines in late 1944, had safely allowed these boats to prowl densely moored minefields and decimate shipping in the Sea of Japan. Refinement of this capability was not aggressively pursued after the war. Sonar improvements were restricted mainly to antisubmarine warfare and limited MCM diver support. In 1945 sonar and primitive electronic underwater locating devices were installed on converted steel-hulled infantry landing craft (LCIL). These redesignated minehunting vessels (AMCU) were originally intended as dedicated minehunters that would locate mines for countermining by divers. Considered ineffective, only three were kept in commission in the Atlantic Fleet in peacetime. Minehunting remained the MCM community's preferred means of actively identifying and neutralizing pressure mines. Still, minehunting sonar developed slowly without priority funding, and by 1950 experimental mine locator sonar had advanced only little.

Progress was made in the longstanding debate between the bureaus over the control of MCM technological development. MCM laboratory staffers had repeatedly made the case for creation of a dedicated MCM laboratory facility within the Bureau of Ships. Finally in 1945 the Navy established the U.S. Navy Mine Countermeasures Station (later Naval Coastal Systems Center, or NCSC) at Panama City, Florida, under the Bureau of Ships. Transferring the ships, equipment, and personnel from the Solomons Island test station to Panama City created a central repository of MCM knowledge and research, a key development in the supporting infrastructure for MCM.[2]

The Bureau of Ships began design work on plans for new MCM vessels. Ships constructed during the war, particularly the capable YMSs, had been built of wood simply for fast production and depended on degaussing of their metal parts to avoid detonating magnetic mines. At Balikpapan, the devastating effects of advanced influence mines on even the carefully degaussed YMSs showed the limits of degaussing as a passive countermeasure. Therefore, although wood became a major feature of postwar minesweeper construction plans, MCM ship design centered on the development of all nonmagnetic parts. By 1949 the Bureau of Ships had produced design characteristics for a new nonmagnetic fleet minesweeper with the operational capabilities of both the converted wooden trawlers and the new steel AMs. Plans for this new AM were shelved due to lack of funding in fiscal year 1950. The Navy did profit, however, in 1946 from the Army's transfer of four small harbor minesweeping boats, the forerunners of the Navy's minesweeping boat (MSB) fleet. Actual funding received for MCM construction went to experimental antipressure mine barges.[3]

Limited U.S. postwar minesweeping activity in the Pacific continued to center only on areas of interest to American vessels, and Seventh Fleet YMSs of Mine Squadron 106 swept contact mines remaining from the war in

Philippine, Korean, and North Chinese harbors where U.S. shipping would be at risk. Similarly, when Chiang Kai-shek requested U.S. transport assistance in removing Chinese Nationalist forces from Indochina in September 1945, the remaining Allied magnetic mines at Haiphong and rumored minefields at the transport area at Do Son required clearance to protect U.S. ships. Japanese minesweeping forces with U.S. advisors were assigned to sweep the shallow and dangerous Haiphong Channel. The YMSs of Mine Squadron 106 swept transport staging areas south of Haiphong at Do Son in October and November 1945 but found no mines. With the addition of some small landing craft (LCVP) with sweep gear, the YMSs joined the sweep at Haiphong and reportedly accounted for five U.S. magnetic mines.

However limited its intentions, the U.S. Navy publicized Minecraft, Pacific Fleet's opening of major Pacific sea lanes to regular peacetime shipping as "the greatest minesweeping job in history." By 1946 the 374 minesweeping ships converted and built for U.S. offensive operations in the Pacific had been reduced to 14 "observation vessels" assisting the Japanese minesweeping effort. Worldwide the Navy retained 37 MCM ships in the active force with 143 laid up in reserve.[4]

Deepening financial problems stalled the Navy's plans for further growth. Just as the United States entered NATO in 1949, committing the nation to continued alliance with wartime European Allies, deep budget cuts hacked away at available resources, forcing the Navy to retrench, particularly in the Pacific Fleet. Worried that communism would spread throughout Southeast Asia, President Harry S. Truman supported French anti-Communist forces in Indochina after 1949 with direct military aid in the form of military advisors and transferred vessels, including minesweepers. The French riverine forces in Indochina relied on active mechanical minesweeping and grappling for controlled mine cables by small boats and on passive MCM to counter mining while at anchor, employing nets, harbor patrols, night watches, lights, and random gunfire and grenade attacks. French involvement in countering Communist mine attacks brought about a resurgence of interest in MCM in the U.S. Navy at the highest levels.[5]

Concern with this Communist offensive mining in Indochina led CNO Admiral Forrest P. Sherman to authorize increased MCM research and development in April 1950. In a report that Sherman approved for action, Navy planners recommended implementation of a complete minehunting system. Minesweeping plans for any immediate conflict centered around employment of three ship types: the World War II DMS for fast fleet sweeping in amphibious assault, the small MSB for harbor clearance, and the newly designed wooden AM for nearly everything else. Based on recent wartime MCM experience, planners recommended improvement of a range of acoustic, electric, and magnetic locating devices; new mechanical and influence sweep equipment; an aerial mine watch system; research on radar and hydrophone use for MCM; increased navigational accuracy for minesweeping vessels; use of both EOD and UDT personnel for mine destruction; improved diver tools; an explosive net sweep; and underwater harbor mapping. Describing minehunting as "the only countermeasure which seems to offer a possibility of being *cheap enough to make peacetime readiness practical*," the report warned that "the great danger is that if mine countermeasures continue to be neglected, large wartime appropriations for countermeasures will be almost useless because the fundamental development will still have to be done first."[6]

Implementation of the report's findings immediately deadlocked in renewed disagreements between the Bureau of Ordnance and the Bureau of Ships over interpretations of jurisdiction. Meanwhile, two concurrent studies of undersea warfare strongly presented similar recommendations for dramatically increasing MCM research, development, funding, education, planning, and coordination. Finding MCM completely "inadequate," Vice Admiral Francis S. Low chronicled the virtual extinction of the wartime mine warfare organization and the abolition of the mineman rating in 1947. "Although later reversed," he reported in April 1950, "this reflected the stagnation which existed in mine warfare until its current emphasis."[7] A Massachusetts Institute of Technology Report on the technical barriers to antisubmarine warfare and MCM also reported that "mine warfare is being pursued in an uncoordinated and unintegrated manner. In consequence, new and unconventional methods are not being explored." To solve the stagnation in MCM research and development, this report recommended unified direction of MCM efforts, proclaiming that "a single organization should have primary cognizance and responsibility for mine warfare in all its aspects, offensive and defensive."[8] A few months later, Sherman would have cause to remember all these recommendations as the inadequacies of the existing mine force became apparent in Korea.

The American-led United Nations response to the invasion of the Republic of Korea (ROK) by the Democratic People's Republic of Korea in June 1950 came under the command of General Douglas MacArthur as Commander in Chief, Far East. Naval forces assigned to him for operations in Korea were commanded by Vice Admiral C. Turner Joy, Commander Naval Forces, Far East. The Seventh Fleet minesweeping force at his disposal when the war began consisted of Mine Squadron 3, made up of six wooden AMSs (redesignated YMSs) and four steel-hulled *Admirable*-class AMs. Three of the latter were in reserve, but readiness of the one in active service was enhanced by its recent employment check sweeping old minefields near Japan. When the Communist offensive began, most of the Seventh Fleet's minesweepers were immediately put on regular escort and picket duty off the Korean coast, and the three in reserve were quickly reactivated. Throughout the entire Pacific Fleet only twelve other minesweepers were active and ready for duty.

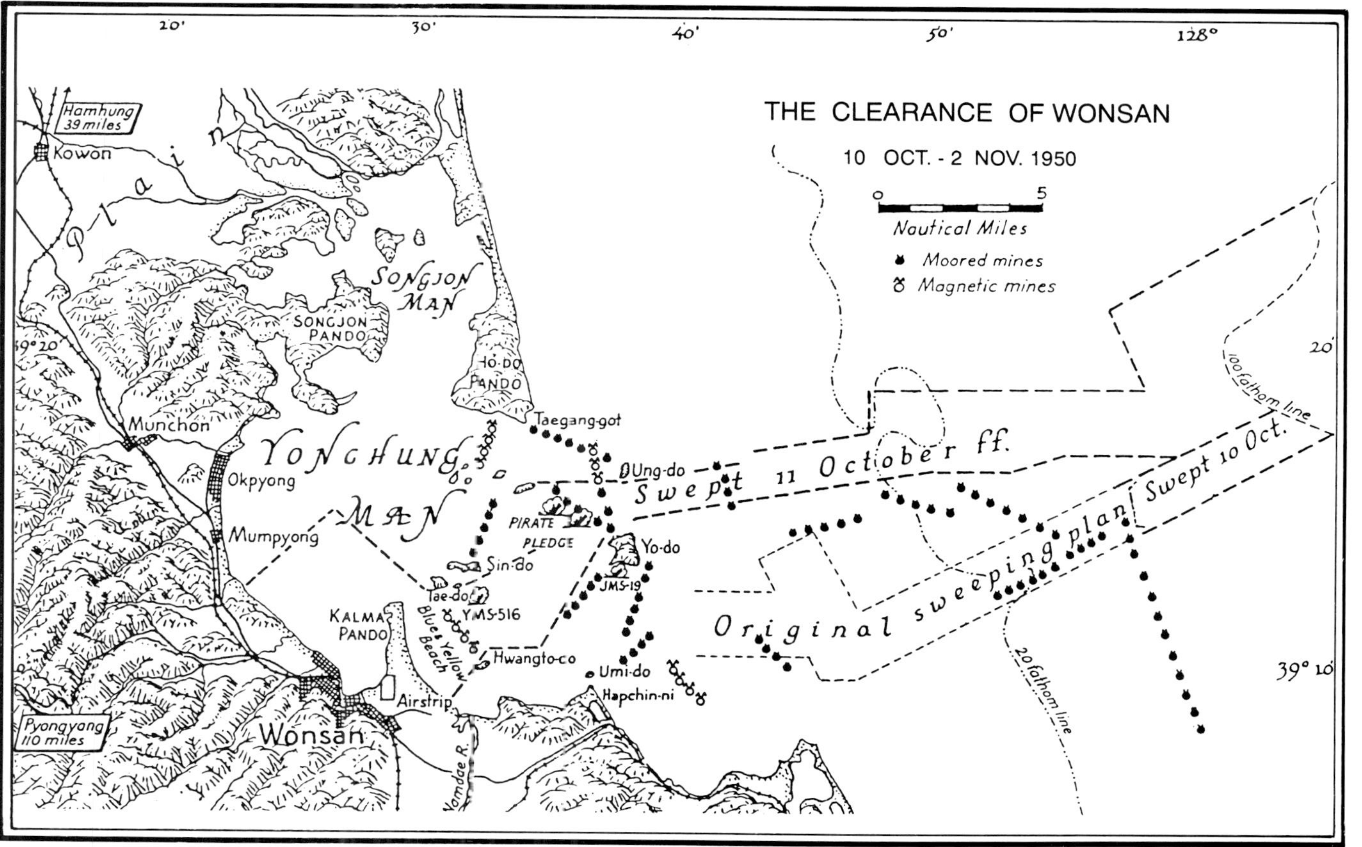

Map of MCM channels in Wonsan harbor. From *History of United States Naval Operations: Korea* by J. Field.

With the outbreak of war MCM was returning to the limelight.[9]

The Soviets had captured nearly all remaining German mines and mine materials at the end of World War II, adding them to their own growing stockpile. More important than the actual numbers of existing mines, however, was the traditional Soviet interest in and development of an effective, professional mine warfare community. The postwar buildup of the Soviet Navy included emphasis on advanced professional education in mine warfare, both for specialists and line officers. The Soviet fleet after 1945 included both wooden- and steel-hulled minesweepers and minehunters with a balanced emphasis on both mining and MCM ability for deep-water, coastal, and riverine operations.[10]

Within weeks of the invasion of South Korea the Soviets sent mines south by rail, undetected by U.N. forces. These included magnetic mines sufficiently sensitive to react to engines and other magnetic parts on wooden ships. Experienced Soviet mine warfare officers personally helped mine the ports of Wonsan and Chinnampo with contact, magnetic, and controlled mines, and instructed North Koreans in mine warfare techniques, dispatching more mines to Inchon, Haeju, Kunsan, and Mokpo.

National Archives 80–G–422837

An LCVP rolls in the well of *Catamount* (LSD–17) during minesweeping operations off Chinnampo, North Korea, November 1950.

American forces in the Pacific quickly became concerned about the possibility of facing mines in their advance. In August 1950 Captain Richard T. Spofford, Commander Mine Squadron 3, warned Vice Admiral Joy that he lacked sufficient vessels and information for any assault sweep. Spofford's concerns were passed on to Sherman and Admiral Arthur W. Radford, Commander in Chief, U.S. Pacific Fleet (CINCPACFLT), who believed, like most officers, that the Korean crisis required the priority reactivation of vessels other than minesweepers.[11]

U.N. forces first found mines in early September 1950 off Chinnampo. Joy's renewed attempts to obtain more minesweepers failed until mines began to take a toll, including a minelayer off Haeju. Sherman immediately deployed several AMSs and DMSs scheduled to arrive in late October, stripping the rest of the Pacific of mine protection. Plans were already in place for a U.N. assault landing at Inchon that could not be delayed. Luckily for the invading force, the North Korean mines protecting Inchon were unsophisticated and relatively few; they were countered without real difficulty. Destroyers in the assault visually spotted and fired on moored contact mines in the channel at low tide. The invading forces passed over the remaining mines at high tide and suffered no mine casualties.

Mines, however, soon did make a difference. In the last week of September mines severely damaged four U.N. vessels and sank one U.S. minesweeper, *Magpie* (AMS–25), on the east coast of Korea, causing the U.N. task groups to operate farther offshore. While looking for *Magpie*'s survivors, a helicopter photographed two mines off Kokoko, demonstrating once again the potential of minehunting by air, which would soon become one of the outstanding developments of Korean operations.[12]

Shortly after retaking Inchon and the capital, Seoul, in September, General MacArthur began planning for a two-pronged invasion of North Korea, with the main U.N. forces advancing overland from Seoul to the North Korean capital at Pyongyang. One corps was assigned to land in an assault on the beaches at Wonsan. MacArthur set D-Day at Wonsan for 20 October, allowing less than three weeks for preparation. Intelligence officers reported several uncoordinated mining attacks and the presence of some influence mines in the general area of Wonsan, but no large minefields were anticipated. With ships in the open Inchon channel beginning to fall victim to newly laid contact and influence mines, however, Joy and his subordinate tactical commander, Vice Admiral Struble, Commander Seventh Fleet, warned their units of the probable mining of Wonsan as Struble's 250-ship landing force steamed toward that harbor.[13]

If anyone knew the limitations of the minesweeping force, it was Struble. As chief of staff for U.S. naval forces in the Normandy invasion, he had seen the effects of influence minefields firsthand; as a task group commander of the 7th Amphibious Force, he had directed assault landings in the

Philippines. Moreover, he was Sharp's successor as Commander Minecraft, Pacific Fleet, and had directed the limited U.S. postwar minesweeping effort in Japanese waters. There were few naval officers with such minesweeping experience. The commander of his minesweeping squadron, Captain Spofford, also had impeccable credentials in mine warfare but they were in ordnance testing and minelaying rather than MCM.[14]

As D-Day approached, Joy, Struble, and their staffs developed reservations about the planned landing at Wonsan and expressed the opinion that the South Korean column advancing rapidly up the east coast of the peninsula would take the port before their amphibious forces arrived. These concerns apparently were never raised with MacArthur, however, so the assault forces proceeded toward their objectives as planned.[15]

Struble sent the Task Force 95 Advance Force, with its Minesweeping Task Group 95.6 under the command of Captain Spofford, toward Wonsan on 6 October. The task group was further increased by Japanese minesweeping vessels. Knowing the Navy's inability to sweep any sophisticated Soviet influence mines that might be in its path, Rear Admiral Arleigh Burke, Deputy Chief of Staff, COMNAVFE, obtained the assistance of twenty Japanese minesweepers for the landing at Wonsan and for cleanup operations at Inchon. After accounting for over nine hundred influence mines in the five years since the end of World War II, the Japanese YMSs, now designated JMSs, were probably the most capable fleet of influence minesweepers in the world.[16] Eight JMSs under the command of Captain Tamura joined the U.N. forces before Wonsan. PBM Mariner patrol planes used to spot mines in the Yellow Sea also augmented this MCM effort.

Admiral Struble personally drew up the minesweeping plans for the operation and took his large task force of three carriers, a battleship, and several cruisers and destroyers to assist the sweep by bombarding the coast. MacArthur had allotted only ten days for the clearance sweep of channels to the beach, and Spofford knew little about conditions in the Wonsan area. Spofford had learned of the minefields at Inchon and Chinnampo and the location of the Soviets' shipping channel; on 9 October he also learned that a U.S. helicopter had spotted a minefield near Wonsan. Knowing that mines awaited him at Wonsan but lacking concrete intelligence, Spofford elected to send his AMSs to sweep a channel directly into the landing area from the 100-fathom curve to meet the deadline for the landing.[17]

The minesweepers began to clear a path on the southern approach toward Wonsan harbor on 10 October, with a helicopter from light cruiser *Worcester* (CL–144) searching the waters in advance. The wooden AMSs, which had never had proper communications equipment installed, received information relayed from the helicopter through *Worcester*. An additional AMS buoyed the swept path in their rear. After sweeping most of the day and accounting for twenty-one contact mines without casualties, the force faced a considerable

Rear Admiral Arleigh Burke, Deputy Chief of Staff, Commander U.S. Naval Forces, Far East, 1951.

NH 50190

challenge when the helicopter spotted five strong lines of mines of an unknown variety in the channel ahead. The minesweepers corroborated the mine lines on sonar and stopped the sweep. Spofford had to choose another channel with fewer mines if the area was to be clear in time for the assault.[18]

Spofford shifted his clearance efforts to the Soviet shipping channel. On 11 October intense minehunting pushed the clearance to the entrance of Wonsan harbor. In preliminary minehunting efforts a PBM spotted and charted minefield locations and alerted the sweepers of the mine positions, while the destroyer transport *Diachenko* (APD–123) sent UDT divers out in small boats to destroy the mines. The following morning the aircraft and frogmen resumed their minehunt while Seventh Fleet aircraft countermined the harbor with 1,000-pound bombs. The arrival of Mine Division 32, commanded by Lieutenant Commander Bruce M. Hyatt, brought the experienced Pacific Fleet steel minesweepers *Pirate* (AM–275), *Pledge* (AM–277), and *Incredible* (AM–249) to assist. With all hands on deck and in life jackets, the division swept forward in echelon.

Helicopters soon reported via radio from destroyer *Endicott* (DD–495) that three lines of mines lay in front of the division, but no position or depth was reported. On board *Pirate*, Hyatt ordered his ships to proceed on course to sweep the newly spotted line of mines. Just as *Pirate*'s bow watch reported a mine contact on the surface, several mines appeared on sonar. Within seconds

National Archives 80–G–421407

Underwater demolition teams from *Diachenko* (APD–123) board a rubber boat that will take them into Wonsan harbor to clear mines, October 1950.

an explosion ripped a hole in the starboard side, breaking the ship in two. In four minutes she went down. Five minutes later *Pledge* was mined. *Incredible*'s engines failed, leaving rescue operations to the wooden vessels. Knowing that Wonsan had already been taken by the ROK land forces on 10 October, seven days in advance of the scheduled landing, and that there was no need to hazard packed troop transports, Struble decided to delay the landing until a path through the magnetic mines could be cleared.[19] With the concurrence of Vice Admiral Joy, Commander Amphibious Task Force Rear Admiral Allan E. "Hoke" Smith informed Admiral Sherman of the mine disaster at Wonsan in terms that left no question concerning Smith's feelings: "We have lost control of the seas to a nation without a Navy, using pre-World War I weapons, laid by vessels that were utilized at the time of the birth of Christ."[20]

Worse news would come. With ninety-two casualties and the loss of the steel minesweepers, surface sweeping halted. For two days PBMs and helicopters carefully spotted contact mine lines while divers hunted, marked, and cleared them, and *Diachenko* searched the harbor with her sonar. Experimental magnetic minesweeping from 14 to 18 October by *Mockingbird* (AMS–27),

Chatterer (AMS–40), *Redhead* (AMS–34), and *Kite* (AMS–22) led to several mine explosions, and one JMS was lost to a contact mine. The channel was nearly cleared when magnetic mines near the shore line destroyed ROKN *YMS–516* on 18 October. Lieutenant Commander Don DeForest, on loan to the task group from Mine Force, Atlantic Fleet, went ashore that day and located some Soviet magnetic coils of the type used to fire these sensitive mines. After interrogating local informants who had helped lay the mines, he returned to the minesweeping group with the basic intelligence on placement and actuation method needed to complete an effective sweep.

Expecting to find relatively few mines in harbor choke points, the task group was stunned over the following week as they slowly charted an extensive mixed minefield of 3,000 mines spread over a 400-square-mile field. Unsophisticated 1904-vintage Russian contact mines in the harbor lay in fields with new magnetic influence mines sensitive enough to react to the wooden ships' engines, making minesweeping by the surface vessels deadly. The line of more sophisticated influence mines close to shore behind a covering

National Archives 80–G–423625

ROK Navy minesweeper *YMS-516* is destroyed after hitting a magnetic mine, October 1950.

Copyright © 1957, U.S. Naval Institute, Annapolis, Maryland.

Influence minesweeping techniques. From *The Sea War in Korea* by M. Cagle and F. Manson.

field of old contact mines caused heavy damage to the AMSs before this pattern was learned. After analysis and application of the information gathered during DeForest's venture, there were no more casualties. Jury-rigging small motor launches and landing craft of the amphibious landing force units to sweep in the shallowest approaches to the beach, Spofford recruited and trained boat crews on site and sent them into the minefield behind the experienced Japanese sweepers.[21]

The actual landing at Wonsan proved anticlimactic after the long delay: fifty thousand men in a powerful 250-ship armada had been held at bay for nearly a week by sea mines. When the U.S. Marines finally landed on the beach, they discovered big ROK banners welcoming them to Korea and a signpost reading, "This Beach is All Yours Through the Courtesy of Mine Squadron Three."

Also on hand to greet them, to their dismay, were U.S. Army Tenth Corps commander Major General Edward M. Almond and Bob Hope, already performing for the men with his USO troupe. "History got ahead of us," noted one commander.[22] Admiral Smith's official report after Wonsan further laid out the feelings of many naval officers, who waited offshore, watching an ill-prepared and under-equipped minesweeping force try to do its job:

> The Navy able to sink an enemy fleet, to defeat aircraft and submarines, to do precision bombing, rocket attack, and gunnery, to support troops ashore and blockade, met a massive 3,000 mine field laid off Wonsan by the Soviet naval experts. . . . The strongest Navy in the world had to remain in the Sea of Japan while a few minesweepers struggled to clear Wonsan.[23]

The humiliation caused by the mines at Wonsan pointed to a critical hole in U.S. naval capability: simple application of extensive fields of Soviet mines, many of them antiques laid by small native vessels, could hold up a superior naval force with inadequate MCM capabilities.

The events at Wonsan had immediate repercussions. Struble's mine force concluded that only an integrated MCM system would provide effective assault clearance in future wars. Struble urged his superiors to develop a sufficient mix of MCM-specific surface vessels, assisted by helicopters to mine spot in the advance, divers to detonate mines, and advanced theater-level intelligence gathering to effect true combat MCM operations and readiness throughout the Navy. He added that "adequate mine countermeasure forces with trained personnel and equipment should be provided in each fleet and should be ready for service."[24] Admiral Joy concluded:

> The main lesson of the Wonsan operation is that no so-called subsidiary branch of the naval service, such as mine warfare, should ever be neglected or relegated to a minor role in the future. Wonsan also taught us that we can be denied freedom of movement to an enemy objective through the intelligent use of mines by an alert foe.[25]

Admiral Sherman agreed that the mining of Wonsan "caught us with our pants down," adding,

> when you can't go *where* you want to, *when* you want to, you haven't got command of the sea. And command of the sea is a rock-bottom foundation of all our war plans. We've been plenty submarine-conscious and air-conscious. Now we're going to start getting mine-conscious—beginning last week.[26]

Mines continued to pose serious problems for the U.S. Navy throughout the Korean War. Sweeping on the west coast continued through 1950 with clearances effected at Kunsan, Haeju, and Kojo. In response to continued North Korean mining, Admiral Sherman immediately ordered the recommissioning of AMSs and AMs as a priority matter. Research into more effective minehunting techniques and equipment began as well. The Navy quickly converted more shallow-draft motor launches to minesweepers and reinstalled World War II sonar, underwater locator equipment, and mechanical and influence sweep gear on amphibious craft. The versatile wooden AMSs continued to be the most able influence sweepers, and the amphibious dock landing ship (LSD) became a capable supply and mother ship for small boats and a logistics ship for helicopters hunting mines.

As the U.N. forces pressed north in October 1950, the heavily mined port of Chinnampo became crucial to resupply the army, and Admiral Joy ordered

National Archives 80–G–423162

A PBM Mariner locates and detonates a mine in the channel near Chinnampo.

Admiral Smith to clear the port of mines. With all Western Pacific minesweepers already fully occupied at Wonsan, Smith sent an intelligence officer to Chinnampo to gather information on the extent and type of mining there and urged the deployment of additional minesweeping vessels from both fleets. Smith appointed Commander Stephen M. Archer to command the sweep operations as Task Element 95.6.9. Recruiting Commander DeForest to assist, Archer commandeered twenty-eight local vessels at Sasebo and threw together a minesweeping force. With a channel to Wonsan cleared, the PBMs also shifted back to the Korean west coast to assist with minehunting.

The mine forces had learned a big lesson at Wonsan: look before you sweep. With that in mind the PBMs and one helicopter spent three days searching for mines at Chinnampo. They identified thirty-four and disposed of most of them by gunfire and depth charges. Although the mines used at Wonsan had proved immune to air countermining, some magnetic mines at Chinnampo were detonated by that means. Surface sweeping began on 29 October by two destroyer minesweepers, *Thompson* (DMS–38) and *Carmick* (DMS–33), later supplemented with three AMSs, two Korean YMSs, and a tank landing ship (LST) with helicopters. After interrogating captured key Korean personnel, intelligence officers uncovered the pattern of a mixed field of 217 contact and 25 magnetic mines by 2 November and the sweep of Chinnampo began in earnest. Minehunting began from the air with planes and helicopters, on the surface with small boats, and underwater with divers. An initial path was cleared around the minefield; this was followed by a second path through the channel entrance. Increased attention to minehunting and good intelligence made the clearance of Chinnampo safe and effective. MCM forces cleared eighty contact mines from two hundred square miles of water without a single casualty. They also proved the effectiveness of the amphibious ships' small boats in sweeping moored mines, the viability of using LSTs as helicopter platforms and supply vessels for the MCM fleet, and the importance of logistics in support of mine clearance.[27]

At Hungnam on 7 November 1950 the mine forces again worked to uncover information concerning the placement of over a hundred moored contact mines. Advance hunting teams of small-boat crews and divers thoroughly scoured the area, attempting to clear a section on the edge of the minefield. AMSs swept carefully for magnetic mines but found none, and the harbor was opened by 11 November. Hungnam sweepers then advanced to open Songjin between 16 and 19 November but found no mines.

MCM units continued sweeping offshore, but changes in the war began to force alterations in MCM application. The approach to offshore MCM during 1950 centered on quickly sweeping clear channels through areas where mines were sparsely laid. For the remainder of the war MCM forces would clear more heavily mined areas to allow close-in gunfire support of land forces by U.N. vessels. By 1951 MCM forces in Korea had been strengthened by more AMSs, AMs, and an LST support ship. More important, the fleet actively coordinated air, surface, and subsurface minehunting forces. In the first weeks of 1951 MCM forces cleared a fire support channel on the east coast

National Archives 19–N–85751

Thompson (DMS–38), shortly after her conversion to a fast assault minesweeper.

near the 38th Parallel, losing *Partridge* (AMS–31) to a mine in the process.

In 1951 U.S. MCM forces continued to face a mine threat in Korea. Near Songjin in the spring, sweepers cleared several contact mines for gunfire-support ships, feinted a landing attempt at Kojo, and check swept off Chinnampo. Enemy remining at Wonsan resulted in more sweeping of that area in March with no casualties. MSBs backed by AMSs again cleared the waters of Wonsan harbor even closer to shore in the summer, disposing of more than 140 mines. Between 1 July and 30 September U.S. minesweepers cleared hundreds of mines from Korean waters, more than had been swept during the entire previous year. In rough November weather at Chongjin, minesweepers faced a tough clearance. The enemy constantly remined in their path, and the helicopters and MSBs had difficulty in the poor conditions.[28]

During the minesweepers' largely successful operations some casualties were caused by mines, particularly by a large number of floaters cut loose from their moorings and drifting with the tide. During another feinted landing at Kojo in late 1952 enemy bombardment took such a toll on the sweepers that they could only complete the sweep at night. Soon nearly all sweeping of Korean waters was done by moonlight, acting on intelligence gathered earlier by daytime air reconnaissance. In addition to performing minesweeping duties, which were reduced to less extensive check sweeps after June 1952, minecraft guarded swept areas, developed mine intelligence methods, and trained South Korean forces in minesweeping and sea-air rescue. By the war's end U.S. forces recorded 1,088 swept mines, all believed to be from Soviet inventory. Still, after the signing of the armistice in July 1953, U.S. vessels continued sweeping and patrolling the west coast of Korea for two months. American commanders at Wonsan respected and appreciated the Japanese minesweeping force that effectively led vulnerable U.S. minesweepers through influence fields.[29]

MCM operations in Korean waters proved the paradox of mine warfare in the U.S. Navy. For more than 140 years the officers and men assigned to MCM had successfully countered mine threats by jury-rigging equipment and by taking measured risks. So successful were their efforts as perceived by the Navy that little funding, prestige, or interest had been given to countering the mine threat, either in war or in peace. Failure at Wonsan changed some of that. During the remainder of the war the inadequate MCM forces succeeded in applying and reapplying the individual lessons learned in the minefields at Wonsan, proving especially at Chinnampo that effective MCM required integration of a variety of surface, air, and subsurface assets. Learning more from failure than success, the generation of Wonsan prepared to stop "damning" the torpedoes and to learn the right lessons from operational experience.

War in Korea and the humiliation at Wonsan helped heighten Cold War interest in naval preparedness against future "Communist aggression," and served to remind the United States of Soviet mine warfare capabilities. Few officers of this generation would forget the hard lesson that fields of even primitive contact mines, sprinkled with a few of the influence sort, could stop the U.S. Navy in its tracks. Nor would they forget that even though minesweeping personnel made up only 2 percent of the naval forces employed in Korea, they accounted for over 20 percent of the naval casualties. MCM readiness was the order of the day. North Korean mines embarrassed the Navy at Wonsan, but they motivated the Navy to make mine warfare and MCM integral parts of the force designed to deal with the growing Soviet threat in the years after Korea.[30]

An immediate lesson learned in Korea was the need for combat readiness of the MCM force. A declining mine force organization after World War II had rendered ships, personnel, and research efforts ineffective. Within the Pacific Fleet the mine warfare type command had been disestablished in 1947. The reestablishment of that command as Mine Force, Pacific Fleet (MINEPAC), at the direction of Admiral Sherman in January 1951 was a positive development in the fortunes of MCM readiness. Subsequently, both fleets for the first time had parallel type commanders responsible solely for the readiness of mine force ships.[31]

The devastating loss of minesweepers to magnetic mines at Wonsan demonstrated the need for a minesweeper design that could nullify the magnetic threat. The Bureau of Ships turned its required design characteristics for a new ocean minesweeper over to famed naval architect Philip L. Rhodes, who had designed several classes of minesweepers for the Navy during World War II, including *Admirable*-class AMs. Rhodes's design for the 172-foot *Aggressive*-class AMs, redesignated ocean minesweeper (MSO) in 1955, had enhanced electrical generating capacity for effective magnetic minesweeping.

Unlike their steel-hulled predecessors that depended solely on degaussing to mask the magnetic parts of the ship, the open-ocean sweeping MSOs were built with wooden hulls, few magnetic materials, and improved degaussing. They were powered by special nonmagnetic Packard or General Motors diesel engines and were also outfitted with controllable-pitch propellers for increased maneuverability in a minefield. Their early UQS–1 minehunting sonar was designed to classify or identify mine types, but this model showed poor resolution and could not classify mines in most waters. They were also rigged with a towing machine on the fantail to stream displacement devices such as the Loch Ness Monster, which never proved effective. Considered the Cadillacs of the international mine fleet of the 1950s, the MSOs were larger than most other wooden vessels and were considered sterling examples of craftsmanship and shipbuilding technology.[32]

In addition to the MSO, the Navy built or converted a variety of vessels along the lines of existing vessels to maximize MCM flexibility. The Soviet mine threat to western European ports led to designs for a new version of the popular YMS, the *Bluebird-* or *Adjutant*-class, 144-foot coastal minesweepers (MSC). The U.S. Navy built 159 MSCs with the new standard controllable-pitch propellers and separate engines to generate electrical power for magnetic minesweeping. All but twenty were transferred to foreign navies under the Military Defense Assistance Program; most went to European NATO allies, who were assigned primary responsibility for Atlantic minesweeping.[33]

From 1952 to 1955 more AMCU minehunters were converted from YMS and patrol craft. The first and only new-construction minehunter, *Bittern* (AMCU, later MHC–43), was based on the MSC. Launched in 1957, it was designed to locate and to plot minefields with a towed array of minehunting equipment and had no neutralization or minesweeping ability. After 1961 *Bittern* was the only minehunter retained in active service.

Other ships designed and converted for MCM operations in the 1950s reflected continued concern for minesweeping components of the future 20-knot amphibious force. The firepower and speed of the fast steel DMSs of World War II were offset by their bulk and vulnerability to influence mines, and by 1955 they reverted to the destroyer force. New plans called for the use of fleets of small, fast boats whose limited signatures would protect them from influence mine detonation and whose size would allow them to be carried in amphibious mother ships. Alterations to the existing MSBs and development of 36-foot minesweeping launches (MSL) were designed to provide a minesweeping capability for shallow waters. MCM ship conversions also included two inshore sweepers (MSI) and two mine countermeasures

USN 1139831

Catskill, the first of the mine countermeasures command ships (MCS), with MSLs on board.

command ships (MCS), LSTs planned in 1956 and commissioned in the 1960s. All MCS configurations carried a coordinated mix of assets: MSBs, MSLs, and an EOD unit for mine disposal. Two Liberty ships were also converted into check and pressure sweepers and designated as special minesweepers (MSS). The importance of the new, visible MCM fleet was reflected in the changes in ship designations. In 1955 auxiliary designations for mineweepers were dropped, and most MCM types received new "M" designators under the special classification for "Mine Warfare Vessels." [34]

The new surface vessels developed after Wonsan were built with sweeper safety as their primary concern. To protect themselves against mines, the ships contained a mix of safety features. These included built-in automatic degaussing systems, nonmagnetic construction, ship silencing, keel-mounted minehunting sonar, and minimal displacement to avoid detonating pressure mines. One important threat to surface sweepers remained: moored mines too deep to be seen from the surface and too shallow to be spotted by sonar. To protect the lead ships in a sweep operation from this threat, testing began to identify existing helicopters for use in a precursor sweep, streaming minesweeping gear.

Among the biggest lessons learned in Korea was the importance of naval intelligence, particularly for accurately ascertaining the configuration and content of minefields. Use of helicopters for mine spotting was an outstanding feature of intelligence gathering on mine positions in Korean operations and a distinct improvement over World War II lighter-than-air craft observations. Helicopters also proved useful for marking minefields with buoys, neutralizing mines, and providing logistic and rescue operation support. [35]

Almost immediately after Wonsan, the MCM laboratory at Panama City developed a separate Air Mine Defense Development Unit and began testing helicopters towing standard minesweeping gear. In early 1952 an HRP–1 helicopter successfully towed *Oropesa* sweep gear to clear contact mines. For thirteen years after Wonsan, as funds sporadically became available, the Panama City MCM laboratory and the Bureau of Aeronautics separately tested helicopters for towing ability, hoping to develop an air MCM (AMCM) capability. [36]

Wonsan broke the fiscal drought on MCM shipbuilding, allowing the development of a more flexible mix of vessels designed to meet the Soviet threat. Before cost overruns on the expanded carrier and guided missile programs forced cutbacks in MCM construction, the Navy had developed an entire program of new ocean, coastal, and small-boat minesweepers and new and converted minehunters, and reactivated several old minesweepers and support vessels. New construction included 65 MSOs, 22 coastal minesweepers, and 1 minehunter, bringing effective strength on paper to 333 vessels, 180 of which were new ships and approximately 93 of which were in active service.

The effects were, however, short-lived. The program was more of an immediate response to the losses at Wonsan than a sustained commitment to the development of mine countermeasures. MCM vessels simply never caught the hearts and minds of most Navy men. After 1958 additional MCM ship construction funds were regularly deleted from the tightening Navy budget, leaving programmed replacements for the MCM force in question for future generations.[37]

Mine research also received priority funding after Wonsan. Countermeasures scientists fought for separate funding, rightly claiming that mine development priorities too often swallowed up scarce resources and left little for MCM. In addition to conducting MCM work at Panama City, the bureaus continued to divide MCM research projects among several laboratories and contractors. Expanded funding and interest in MCM problems within the scientific community after Wonsan, however, facilitated cooperation and the intensive networking required to achieve results within the Navy's decentralized organization.

In 1951 the Office of Naval Research contracted with the Catholic University of America to form the Mine Advisory Committee of the National Academy of Sciences, a study group of academics organized to investigate mine and MCM technology. Because of Wonsan their first nineteen years of work were devoted to MCM studies, including Project Monte (1957) on countermeasures research, the Precise Navigation Project (1966), the High Resolution Sonar Project (1968), and Project Nimrod (1967–1970), the latter an analysis of the present and future of mine warfare. Developed out of a summer research group at the Naval Postgraduate School at Monterey, California, Project Monte surveyed MCM development and made recommendations for advancement. From this project several MCM research plans were developed, including small manned or remotely operated underwater vehicles to sweep, hunt, classify, and neutralize mines in advance of surface ships.

Panama City expanded its MCM programs, particularly in testing helicopter minesweeping, but other laboratories also developed MCM technology as an adjunct to their missions. In the 1950s the Navy Electronics Laboratory began developing continuous transmission, frequency-modulated sonar units that had been used in World War II submarines into portable sonar units for divers. Scientists at Yale and NOL examined traditional and nontraditional methods for minesweeping, hunting, identification, and neutralization and encouraged mine warfare and countermeasures studies through yearly conferences. Operational testing within the fleet developed experimental air cushion vehicle technology as protected minesweepers. After 1957 operational problems were addressed by NOL participants in the Naval Science Assistance Program, a Navy-wide program designed to provide direct scientific support for fleet needs.[38]

Scientists at the various laboratories expanded earlier attempts to develop countermeasures against increasingly advanced combination influence mines. Attempts to design a vessel of sufficient size to actuate mines yet withstand the shock of the explosion had varying degrees of success.

The XMAP (Experimental Magnetic, Acoustic, and Pressure Sweep) was a 250-foot-long, 19-foot-wide, 2880-ton, tightly-welded steel cylinder device towed by tugs and designed to counter all influences in one sweep. Originally it had been a 1944 design proposed by scientists at the David Taylor Model Basin. Once approved for development after Wonsan, XMAP was almost immediately embarrassing because of massive cost overruns. After near-abandonment of the scheme, XMAP–1 was finally completed and towed to Panama City for testing in 1956. The test results were never released, but XMAP reportedly suffered from maneuverability problems and may have failed to provide sufficient pressure signature to fire mines. In 1961 it was slated for disposal.[39]

New designs for unsinkable ships proved similarly unsatisfactory, costly, and impractical. As a last resort, plans for displacement devices again centered on guinea pig ships as check sweeps, leading to experiments with enhanced buoyancy, reverse degaussing, and topside-control for large, expendable vessels. Aside from those countered by stop-gap measures, pressure mines remained virtually unsweepable.[40]

Minehunting research focused on the use of sonar or magnetic anomaly detectors to locate mines. Working on earlier sonar studies of the Navy Electronics Laboratory, the Harvard Underwater Sound Laboratory, and the General Electric Company, scientists at the Applied Research Laboratory of the University of Texas at Austin developed an improved, high-definition, variable-depth AN/SQQ–14 Mine Detector/Classifier sonar to identify mines for neutralization by EOD divers. This sonar replaced the UQS–1 sonar on MSOs and became the worldwide standard for MCM minehunting and classification, effectively isolating mines from other debris littering harbor floors. Meanwhile, MCM scientists applied evolving precise navigation and position fixing to MCM to produce safer, more accurate, and more even sweeps.[41]

Attempts by the Department of Defense to create parallel lines of research authority in all the services mandated changes in Navy research organizations, further affecting MCM research and development. In 1965 and 1966, under pressure from DOD, the Navy detached the laboratories from the bureaus, having them report to a single Director of Naval Laboratories and ultimately to the Chief of Naval Material. This restructuring did not, however, fully resolve the issue of jurisdiction over MCM studies. Although the laboratories now reported to the Director of Naval Laboratories instead of to the parent systems commands that had replaced the old bureaus, scientists who paid attention to the technical history of MCM discovered that various laboratories, all claiming cognizance, had in some cases been studying and restudying the same methods of scientific MCM, unaware that some of the approaches had already failed.

Funding problems also began to affect the research community in 1965. In that year the Navy built the last surface sweeper of the Wonsan shipbuilding program funded in fiscal year 1958, and production of advanced MCM sensors and other equipment already designed was delayed by conflicting funding priorities.[42]

The most important gain made by the MCM program after Wonsan was in the availability of qualified people. The reinstatement of MINEPAC in 1951 and the arrival of new ships in the fleet after 1953 increased the needs of the mine force for both MCM experts and surface ship commanders. For a brief period in the late 1950s MCM had the potential to become a viable professional path for some career officers. MSOs came off the assembly line regularly from 1952 to 1956, and recruitment of crews to man the new ships made training of MCM personnel a naval priority. The Naval Postgraduate School offered a two-year master's degree in mine warfare from 1955 to 1960 (although only twenty-two officers completed the program), and Yorktown's Mine Warfare School also improved its MCM courses for officers and enlisted men. With active duty officer and enlisted billets available in the operational MCM fleet and support facilities, the opportunity to build a stable, active duty MCM community existed for the first time.[43]

MINEPAC headquarters centered many West Coast minesweepers at Long Beach, California. Combat readiness required extensive exercise of the new ships and people, so MINEPAC minesweepers deployed on yearly six-month tours in the Western Pacific. MSO divisions in MINELANT, regularly deployed in the Caribbean and the Mediterranean on similar tours as "ready" units, and returned home to either Little Creek, Virginia, or to MINELANT headquarters at Charleston, South Carolina. In 1959 the Mine Warfare School moved from Yorktown, Virginia, to Charleston, further establishing that city as a home for the mine warfare community of the Atlantic Fleet. Extensive shore maintenance and support activities for the MCM force were also established at Charleston and Long Beach with some support facilities at Key West, Guam, and Sasebo.[44]

For young officers and enlisted men in the late 1950s and early 1960s, assignment to the new MCM force provided an unusual experience in both seamanship and leadership. Command came early, and the career advancement possible with MCM ship command enticed some of the most promising graduates of the destroyer force schools into the new mine force for at least one command tour. Young lieutenants obtained command of MSCs; lieutenants and lieutenant commanders captained MSOs; ensigns served early tours as department heads; and lieutenants (junior grade) served as executive officers. Senior enlisted men who commanded MSBs and smaller

U.S. Navy Photograph

Dash (MSO–428), one of the *Agile*-class ocean minesweepers built in the mid-1950s.

vessels often advanced into the MCM officer community through such experience. Because the establishment of minesweeping divisions, squadrons, and flotillas provided MCM billets for commanders and captains, and because of the variety of MCM vessels, shore station assignments, and missions, it was actually possible for a brief time for an officer or an enlisted man to rise within the mine force to the rank of captain.

Everything that had to be done on a big ship also had to be done on a small one, and the expanded MCM force became a hands-on training school for a whole generation of naval officers who exercised command at an early age. Officers assigned to the MSCs and MSOs from the active duty destroyer force sometimes arrived with little or no training in mine warfare and began operating immediately. Junior officers, many of them ensigns right out of school, often had good technical training from the mine warfare school but lacked basic shipboard experience. Well-trained enlisted men, both active duty and reserves, made up the core of the MCM force and usually taught their officers the essentials of minesweeping and hunting on the spot.

Minesweepers often operated in close sweep formation with other vessels of their units, and thus had requirements for precision navigation and seamanship that were well beyond those of most larger ships. MSO divisions of four or five ships, supported by an amphibious mother ship, swept staging areas and channels in fleet exercises with both mechanical and influence minesweeping gear and, in formation or singly, practiced using sweep gear, minehunting equipment, and EOD teams in tactics appropriate to each situation. Because of the exacting navigation needs of minesweeping vessels, crews gained great experience in piloting. Few destroyer force officers who passed through the MCM force ever swept a real mine, but they gained early command experience and, perhaps more important, familiarity with the more practical aspects of mines and MCM.[45]

The generation of officers that passed through the mine force at its highest point of funding after Wonsan saw a navy in microcosm: a small, vigorously active, and close-knit community that suffered from the same problems experienced by the Navy as a whole—plus a few additional ones. The MSOs had serious difficulties with the brittleness and unreliability of their aluminum and stainless steel nonmagnetic diesel engines, particularly the Packard types; those built with nonmagnetic General Motors engines had fewer failures. The small ships were also labor intensive, and their wooden construction added to the hazards of fire. Most damning, they had only been designed for a top speed of fourteen knots, and slow-but-capable did not fit the needs of the faster, forward-deployed postwar Navy.[46]

The growth of the MCM community reached its peak in the decade after Wonsan. Although increased assets briefly made MCM a more dependable career specialty with many opportunities, those assets could and did go away. As the MCM community struggled to develop combat effectiveness, circumstances simultaneously began to chip away at the position and priority MCM had achieved within the Navy. Although Wonsan made the entire Navy briefly more mine-conscious, competing concerns quickly returned MCM to its isolated position.

"Neither that war," noted one observer after Korea, "nor more recent developments have made a career in mine warfare a realistic or attractive alternative to the many ways in which an officer can rise in the Navy."[47] Those recent developments would permanently alter the focus of future MCM, for after 1965, U.S. involvement in Vietnam would drain the Navy of its assets, and the MCM force of the Wonsan generation would not be the only casualty.

4

New Lessons Learned:

The Impact of Low Intensity Mine Warfare
1965–1991

The U.S. naval MCM experience in Vietnam from 1965 to 1973 significantly altered the Navy's perception of MCM as an integrated element of naval warfighting. Mines and MCM played little part in the daily operations of most naval units operating in the coastal war offshore. Important shallow-water MCM operations on the rivers of Vietnam, however, required close coordination with river patrol and special operations forces. This situation slowly altered the shipboard Navy's perceptions of the character of MCM over the course of the war. By the time a substantial mine offensive required commitment of a large-scale MCM operation, the Navy had come to view MCM as a small-scale specialty rather than as a major element of naval warfare.

After the Communist Viet Minh defeated the French in Indochina in 1954, the resulting Geneva Agreement divided Vietnam into two halves at the 17th Parallel, or Demilitarized Zone (DMZ), separating Communist forces in the north from the non-Communist forces in the south. From 1954 to 1959 U.S. Navy personnel assigned to the Military Assistance Advisory Group, Vietnam, worked to develop a fledgling South Vietnamese Navy, providing MCM advice, vessels, and training. As the North Vietnamese worked to extend Communist control over South Vietnam and neighboring Laos from 1959 to 1961, U.S. Navy carrier task forces were deployed off the Vietnamese coast to deter further encroachment. Seventh Fleet MSOs participated in a minor way, establishing an American presence in Indochina by making port visits in Cambodia in August 1961.[1]

Later that year the Kennedy administration responded to escalating Communist insurgency with U.S. ships and men. With limited funds available for new patrol craft, ocean minesweepers were tasked with collateral patrol duty. In December 1961 the MSOs of Mine Division 73 and later of Mine Division 71 patrolled near the 17th Parallel to stop or deter North Vietnamese coastal smuggling of arms into South Vietnam. In these patrols the MSOs joined U.S. Navy destroyers and used their radar to vector Vietnamese Navy ships to suspicious vessels. Finding little evidence of North Vietnamese infiltration from the sea, the MSOs ceased patrolling the following August. When President Lyndon B. Johnson increased U.S. counterinsurgency support to South Vietnam in 1965, thus escalating American military involvement, U.S. naval forces joined the regular South Vietnamese Navy coastal antiinfiltration patrol in an operation designated Market Time. MSOs and MSCs again shelved their sweep gear and went on patrol duty.[2]

Actual MCM experience for the Vietnam MSO fleet was rare. In January 1968 the MSOs of Mine Division 91 were pulled off Market Time patrols to clear what was believed to be the first known live minefield since 1953, created when, under unexplained circumstances, U.S. forces lost a load of mines in the Tonkin Gulf. After its attempts at sonar minehunting failed because of the heavy mine-like litter on the ocean floor, the division necessarily reverted to mechanical and influence minesweeping. Since the North Vietnamese did not mine their coastline, MCM was never part of the regular Market Time scenario. Tasked primarily with ancillary duties, the MSO fleet that was the product of Wonsan would be a minor player in coastal warfare. The real mine battle would be inland, where the U.S. had rarely prepared to fight.[3]

Like the Union forces of the Civil War era to which they were often compared and the French Navy from which they inherited the war, the U.S. riverine forces faced a shallow-water MCM challenge requiring ingenuity and intraservice cooperation. The Navy's traditional preoccupation with oceangoing ships and major amphibious assault landings from the sea faced a new challenge in the rivers of Vietnam.

Like the Confederate Navy, the North Vietnamese relied heavily on mines to attack ships in the rivers, and often coordinated mining with gunfire and rocket attacks. Viet Cong mines ran the gamut from simple contact mines to a few advanced Soviet influence types, but most often U.S. Navy MCM forces faced homemade controlled mines (usually waterproofed land mines with command detonating cables), drifting mines disguised as garbage, and swimmer-delivered limpet mines that adhered directly to boat hulls. Countering them required hard work, technical expertise, and the ability to jury-rig equipment.[4]

French Navy river assault divisions conducting counterinsurgency patrols in 1953 and 1954 had employed tugs and mechanized landing craft (LCM) as minesweepers, cutting controlled mine cables with a drag sweep. Eventually supplementing their minesweepers with air-spotting and ground support to protect harassed river units and to overrun command detonating centers, the French had learned that the best countermeasure on the Indochina rivers was cooperation among all air, patrol boat, and ground forces to prevent mine planting and attack.

In October 1965 when shallow-draft U.S. Navy MSBs began operating in the rivers of South Vietnam, they learned the same lesson. Originally designed to operate in water under the control of friendly forces, MSBs were adapted to sweep under close-in combat conditions by adding extra armor in the form of fiberglass resin or ceramic material to retain their nonmagnetic signature. Additional armament included a .50-caliber machine gun.

A minesweeping boat (MSB) patrols the Long Tau River to keep the channel safe for commercial traffic moving into Saigon.

Dragging hooks attached to a winch by steel wire to cut electrical controlling cables, MSBs became a favorite target of the Viet Cong, who continually harassed them with sniper fire. Effective mine clearance on the rivers of Vietnam once again required MCM vessels to operate jointly with other river and patrol forces for protection.[5]

Riverine warfare, originally additional duty of the Market Time forces, produced two separate river patrol operations: Task Force (TF) 116, River Patrol Force, established in 1965 to interdict enemy traffic on the major rivers of the southern delta, and Task Force 117, Riverine Assault Force, an amphibious assault force created in 1966 in response to increased mining and guerrilla attacks. In the Mekong Delta region Task Force 116, a major participant in Operation Game Warden, began an extensive patrol effort to control access to the complex river system south of Saigon in December 1965, particularly in the Rung Sat Special Zone, a Viet Cong base. Of particular importance to Task Force 116 operations was the forty-five miles of waterway of the Long Tau River, the main shipping channel to Saigon and a key logistic point for the U.S. forces. Heavy mining of the Long Tau with both Soviet mines and homemade Viet Cong controlled mines increased the need for minesweeping activity south of Saigon.[6]

Four MSBs operating near Nha Be on the Long Tau participated in Operation Jackstay in March 1966, a quick strike operation designed to clear Viet Cong guerrillas from their stronghold in the Rung Sat. Two months later the Rung Sat River Patrol Group established a base at Nha Be for river patrol boats (PBR), large personnel landing craft (LCPL), and detachment Alfa of Mine Squadron 11 (later Mine Division 112), consisting of eight MSBs with mechanical sweep gear, two officers, and a hundred enlisted men. Although the first mines found by the Nha Be forces were controlled and limpet mines with timers, in December 1966 an MSB swept and recovered a 2,000-pound Soviet contact mine.[7]

As mine attacks increased, the Nha Be detachment added four additional MSBs and LCMs rigged with standard minesweeping gear and redesignated river minesweepers (MSM). Operating daily to keep the rivers south of Saigon clear of mines and targeted by Viet Cong units, the MSBs of Nha Be suffered severe damage. Heavily armed with machine guns and grenade launchers, Nha Be forces swept for moored contact mines and controlled mine cables and countered swimmers carrying limpet mines with hand grenades. This duty, at times the only active U.S. Navy minesweeping effort, led them throughout the dangerous Rung Sat, sweeping south to the ocean from their base while South Vietnamese Navy minesweeping launches swept north to Saigon. Like the French before them, the Nha Be minesweepers found that effective MCM against guerrilla units required coordinated effort by all U.S. forces operating nearby.[8]

As cooperative mining and guerrilla attacks on U.S. ships in the Rung Sat increased, U.S. forces prepared a combined response. In support of Task Force 116's increasing patrols, Task Force 117 began operating a mobile base and assault force to counter Viet Cong attacks through minesweeping, land and river patrols, and river bank defoliation, which materially assisted in preventing controlled mine attacks. Two mobile support bases for the joint forces joined Game Warden in 1967. Electronic detection devices and strong currents protected these pontoon barge bases from swimmer attacks, but appearance of these devices on the rivers caused an escalation of mine attacks and ambushes of patrols. Game Warden minesweeping units on the rivers of South Vietnam cooperating with other Navy and Army units averaged approximately seventy-five patrols per month. Mobile bases for the river force included self-propelled support ships and non-self-propelled platforms—moved by tugs and outboard equipment—such as the repair, berthing and messing barge (YRBM), units of "married" barracks craft (APL), and floating workshops (YR) that provided logistic support for MCM operations.[9]

In the northern provinces additional minesweeping units were established at Da Nang and Cam Ranh Bay in April 1967. These units swept the Cua Viet River near the DMZ and the Perfume River, which ran to Hue, to combat regular mining by the Viet Cong. After the 1968 Tet Offensive the Viet Cong increased mining of the Cua Viet, often using drifting mines disguised in baskets; U.S. forces countered these with hand grenades. LCMs rigged with minesweeping gear joined the PBRs assigned to the river patrols of Task Force Clearwater, protecting supply routes on the Perfume and Cua Viet rivers and providing convoy protection. North Vietnamese mining of the Cua Viet intensified in early 1969, and Mine Division 113 sent three MSBs to assist. By 1970 four minesweeper monitors (MSM) were regularly assigned to sweep the Perfume River while five went to the Cua Viet as part of the river security groups. At least one pressure mine was used on the Long Tau and one on the Cua Viet, where two magnetic-acoustic mines were also discovered.[10]

Riverine warfare in Vietnam required a variety of small boats to meet the mine threat. In addition to MSBs, the river forces converted other craft to minesweeping duty. The 30-ton patrol minesweeper (MSR) and the MSM swept both mechanically and acoustically. Other vessels including a modified 50-foot motor launch (MLMS) and the 36-foot plastic MSLs deployed from MCM support ship Epping Forest (MCS–7) swept for combination magnetic-acoustic mines in the Cua Viet River in 1968.[11]

In response to the needs of riverine MCM, the Panama City scientists developed magnetic and acoustic sweeps, infrared searchlights, a chain drag sweep with cutters to catch controlled mine cables, and improved sonar for testing in the rivers of Vietnam. They also developed experimental drone minesweepers (MSD), 23-foot, unmanned remote-controlled boats designed to sweep in shallow waters. Mine Division 113 tested the drones after February 1969 in the Mekong Delta, but they failed to clear mines and were later transferred to the Vietnamese. Most often the simplest methods worked best. MCM forces mechanically dragged and swept for control cables and contact mines using the new gear designed by Panama City. To counter swimmers with limpet mines, ships and barges used nets, patrols, regular watches, and randomly lobbed hand grenades to discourage swimmer approach.[12]

Riverine MCM required adaption of MCM technology to brown-water craft, while the surface MCM ships of the blue-water Navy operated continuously in the collateral duty of coastal patrol. By the late 1960s the MSO ships, the backbone MCM platforms, were in serious disrepair. Their brittle nonmagnetic engines and wooden hulls had been designed for the most dangerous magnetic minefield conditions, not for the constant patrols and escort operations that exigency now demanded. The high costs of Vietnam operations forestalled plans for a 1965–1966 minesweeper design to replace the MSOs. The proposed new sweeper was designed to carry the Shadowgraph side-looking mine detector sonar and a wire-guided torpedo known as the "Sea Nettle" for mine disposal, but the cost of operations depleted all ship development funds. Instead, the Navy had to find a means to keep the MCM

U.S. Navy Photograph

A sailor prepares to lower a drag chain with cutters into the Long Tau River.

fleet, like other aging naval vessels, operational.[13]

Block obsolescence, a continuing problem for all ships of the U.S. Navy since World War II, required rehabilitation programs to extend service life expectancies. The capable MSOs, plagued with persistent mechanical problems and suffering from excessive wear on their wooden hulls, required significant upkeep, and in the late 1960s an MSO modernization program was begun. Improvements included installation of new Waukesha nonmagnetic engines and the SQQ–14 minehunting and classifying sonar, which significantly enhanced the ships' performance. Unfortunately, uncontrolled cost growth, technical deficiencies, and the lack of a long-term, stable production program designed to meet the special needs of wooden ships and minesweeping requirements resulted in an embarrassing fiasco. The modernization attempt reached its peak in 1969 with only thirteen of sixty-five ships completed, just when shipbuilding and refitting funds were being diverted to East Asian operations. Faced with expanded operations, plummeting resources, and a failed modernization program, the surface MCM force seemed doomed to extinction in a generally worn-out Navy.

Escalation of the Vietnam War and increasing fiscal austerity through the late 1960s started the MCM forces' long descent. Continuous commitment of U.S. naval ships and personnel around the globe and in Vietnam drained all Navy resources for the length of the war. Many ships served a dual purpose; the patrol and surveillance missions assigned to minesweepers freed up other Navy assets but shortened the life span of these rapidly aging vessels. Limited funds, stretched to fulfill wartime operational and logistic needs, left little money for completion of the refurbishment program. Ships that wore out quickly were not replaced, reducing the number of MCM ships and the billets assigned to them. The remaining ships and slow-moving river assault forces seemed plodding and outdated by 1970, particularly when compared with the quick-strike capabilities of readily available helicopters.[14]

One of the key lessons of Wonsan had been recognition of the U.S. Navy's limited shallow-water MCM capabilities. Although mine warfare in the rivers of Vietnam reminded the rest of the Navy of that lesson, the mine warfare community had never forgotten it. Intrigued by the possibility of sweeping by air to protect MCM surface vessels from mines, the mine force's interest in AMCM increased during the war. Operational helicopter experience in Vietnam also taught the MCM force that protection was a two-way street: in any amphibious assault requiring a precursor sweep, AMCM units would require at a minimum escort by offensive gunships and would prefer sweep areas not subject to hostile fire. Developmental AMCM targeted the MCM needs that surface vessels could not meet, particularly protection of the lead sweeping vessels from sensitive mines.[15]

Despite limited funding and support for AMCM development, laboratory work had steadily advanced. Panama City had identified and tested helicopter types suitable for minesweeping and had continued experimenting with sweep and minehunting gear. In the mid-1960s Panama City and the Naval Ordnance Laboratory collaborated on a project called "turtle," a helicopter-towed underwater mine detection system. The turtle-shaped unit included a small sonar unit, a television camera, and explosives to hunt, classify, and if possible detonate mines from a distance; however, the poor quality of the television picture and inadequate maneuverability shelved this project indefinitely. Panama City also began development of a magnetic sweep sled, a hydrofoil vehicle housing an influence generator capable of high-speed sweeps.[16]

Meanwhile, Mine Squadrons 4 and 8 at Charleston experimented with helicopters and new gear throughout the late 1960s to develop tactics and support for precursor mechanical minesweeping by helicopter. After adapting mechanical sweep gear used by MSBs and MSLs for the helicopters, the squadrons tried to sweep by towing MSLs generating their streamed magnetic sweep gear, but this kind of activity required a boat crew.

The squadrons soon found there were certain limits to using helicopters as minesweepers. Helicopters are not autonomous vehicles. They require a base ship and support vessels for operations, particularly to deploy and retrieve the heavy sweep sleds. Traveling at relatively slow towing speeds without engine overheating was an early problem, and helicopter rotor noise set off sensitive mines. In addition to support ships and a substantial logistics chain to operate, the helicopters also required good weather and atmospheric conditions. Nonetheless, by 1970 a promising start had been made in mechanical clearance by helicopters.[17]

When Admiral Elmo R. Zumwalt, Jr., became Chief of Naval Operations in 1970, he declared his immediate intent to rescue mine warfare from its obvious decline within the Navy, despite drastic cutbacks in DOD funding. Setting himself up as the champion of the mine warfare program, Zumwalt explained:

> I think that we in the U.S. Navy ... have frequently been accused of not giving sufficient interest to the field of mine warfare, and in part, I believe, this is the result of the fact that our Navy is made up of three unions: the Aviation union, the Submarine union, and the Destroyer [surface] union and I have therefore made myself the head of the mine warfare union to try to get an equal balance of interest within the United States Navy in this very important field.[18]

As a young lieutenant in November 1945, Zumwalt had witnessed Pacific minesweepers with inadequate intelligence and minehunting capabilities struggling to clear uncharted minefields at the mouth of the Yangtze River. He concluded at the time that surface minesweeping vessels were slow and uncertain. By 1970 he also considered them outdated. Within sixty days of becoming CNO, Zumwalt had developed Project 60, a comprehensive plan to revitalize the U.S. Navy during his tenure, and had decided to push through a complete helicopter MCM program. Believing that the aging surface MCM force was a financial drain on scarce Navy resources, Zumwalt scrapped the surface MCM fleet that was the product of the lessons learned at Wonsan "to economize on and modernize minesweeping techniques." He predicted huge savings in operating expenses and the development of a worldwide, quick-strike, cost-effective, and safe method of countering the Soviet mine threat. The only problem was that the systems for using the helicopter as the sole sweeping unit were still developmental.[19]

Zumwalt's plan effectively reversed the trend of MCM development since

1952. Tactics and equipment had been developed to use helicopters to sweep sensitive and shallow mines in advance of regular surface minesweepers and minehunters; now the remaining surface ships would support the helicopters. Zumwalt gambled that the aircraft could be procured, tactics developed, and personnel trained before any mines needed to be swept.[20]

These MCM policy changes reorganized MCM assets. As Vietnam naval operations permitted, MSOs stopped deploying with the fleets. The MCS mother ships were decommissioned, and helicopters and MSLs were embarked as needed on other amphibious ships. MSCs and some MSOs were sold to allied navies or transferred to the reserves, and only the MSBs and a few MSOs were retained to train the active fleet for wartime contingency operations. Impressed by Zumwalt's belief in replacement of the surface MCM force with AMCM assets, the Department of Defense pressed for cancellation of a new 1971–1973 building program of MSOs and MSCs. All remaining funding and planning centered on the new aviation technology.[21]

Zumwalt recognized the need for centralized control of the diverse air, sea, and undersea MCM assets. On 1 July 1971 he consolidated MINELANT and MINEPAC under one two-star type commander, as Commander Mine Warfare Force (COMINEWARFOR), at Charleston. COMINEWARFOR reported administratively to the CNO through the fleet commanders and operationally acted as a task force commander assigned to fleets as needed. Responsible for the readiness of all MCM units, COMINEWARFOR operated and maintained all surface MCM units and AMCM sweeping systems, with air assets actually owned and administered by Commander Naval Air Force, U.S. Atlantic Fleet (COMNAVAIRLANT). He also held sole responsibility for readiness and minefield planning. Active and reserve minesweepers reported "operationally, administratively, logistically, and training-wise" to COMINEWARFOR. For the first time since Admiral Sharp took over Minecraft, Pacific Fleet, during World War II, one officer exercised command responsibility for operational readiness of all MCM assets.[22]

With establishment of centralized coordination for the MCM forces, the next priority was procurement of AMCM aircraft. The Marine Corps agreed to transfer fifteen CH–53A helicopters strengthened with tow points to the Navy and also accepted a secondary MCM mission for their own CH–53A pilots and aircraft. COMNAVAIRLANT took thirteen of the transferred Marine Corps CH–53s and established the first operational AMCM squadron, Helicopter Mine Countermeasures Squadron 12 (HM–12) at Norfolk, Virginia. COMINEWARFOR integrated this squadron into MCM operations by reestablishing Mine Squadron 8 as the Mobile Mine Countermeasures Command (MOMCOM). The MOMCOM's mission was to train both MCM and AMCM assets for operations and to develop surface and airborne tactics and command and control capabilities. As additional duty, the MOMCOM commander would also deploy on short notice as an operational MCM task

Courtesy CAPT Paul L. Gruendl, USN (Ret.)

MSB–40 trains with magnetic sweep gear before Operation End Sweep.

group commander. His mission was to develop a global AMCM unit capable of deploying within seventy-two hours and of independently sweeping contact and influence mines for up to ninety days. MOMCOM would control all equipment, training, sweeping, and air control techniques, while COMINEWARFOR would coordinate, acquire, and deploy the air and surface units.[23]

Together, HM–12 and MOMCOM began practicing streaming, towing, refueling, and retrieving sweep gear and developing design modifications for a dedicated minesweeping helicopter in April 1971. The first unit of the new AMCM program underwent its first overseas exercise in a Mediterranean deployment in October 1971, and by February 1972 a second AMCM unit successfully participated in an amphibious assault landing exercise off the coast of Maine. By April a third exercise, a breakout channel clearance sweep at San Diego, California, was completed.

Things moved quickly in the AMCM effort because they had to. The war in Vietnam was dragging on and MINEWARFOR staffers were drawing up

100

plans for the mining of the waters of North Vietnam, particularly the crucial port of Haiphong.[24]

Despite U.S. efforts to disengage from Vietnam, once the North Vietnamese began their concentrated March 1972 Easter Offensive on South Vietnam, the Seventh Fleet returned in strength offshore. With peace talks in Paris paralyzed, President Richard Nixon ordered a full commitment of U.S. naval forces to blunt the North Vietnamese assault. The fleet, together with the Air Force, launched a massive bombing campaign against North Vietnam.

Nixon also ordered CINCPACFLT Admiral Bernard A. Clarey and Commander Seventh Fleet Vice Admiral James L. Holloway III to prepare plans to mine Haiphong. Knowing that the Hague Convention of 1907 required subsequent removal of the mine threat and that the North Vietnamese would demand it, MINEWARFOR was tasked to prepare to sweep the mines soon to be planted. Consequently, planners from the Seventh Fleet and MINEWARFOR who developed the minelaying campaign also prepared in advance for the mine clearance, stipulating mine types requiring only magnetic sweeps for clearance. "From the beginning," one officer recalled,

> the possibility of U.S. forces having to sweep the mines was a factor which influenced the types of mines used, their settings, and to a lesser degree their locations. As a result, when it came time to sweep, we knew everything about the mines and had purposely planted mines which could be swept easily and effectively by our mine countermeasures forces. . . . The vast majority of the mines were programmed to self-destruct and the remainder to go inert after a given time. Thus, even as the mines were dropped, the process of mine removal had been started.[25]

On 8 May 1972 attack aircraft from *Coral Sea* (CVA–43) began mining North Vietnam's major ports. The first drop in Haiphong harbor consisted of thirty-six magnetic-acoustic mines, which immediately stopped virtually all ship traffic. The North Vietnamese did not know how many were dropped and made no immediate attempt to sweep them. Seventy-two hours later the mines armed themselves, sealing twenty-seven foreign merchant vessels in the port. President Nixon announced that the mines would not be removed until the release of all American prisoners of war. Continued remining and bombing of North Vietnam influenced negotiations in Paris through 1972 as the U.S. increased military pressure on the North Vietnamese to negotiate a settlement.[26]

In September 1972 Rear Admiral Brian McCauley was ordered to report as COMINEWARFOR and for additional duty as Commander Task Force 78, Mine Countermeasures Force, U.S. Seventh Fleet. Although Admiral McCauley had no mine warfare experience, his staff and operational officers did. Captain Felix S. "Hap" Vecchione, commander of the versatile MOMCOM and the driving force behind the study of operational MCM, directed MCM task group operations. His two key AMCM detachments were led by

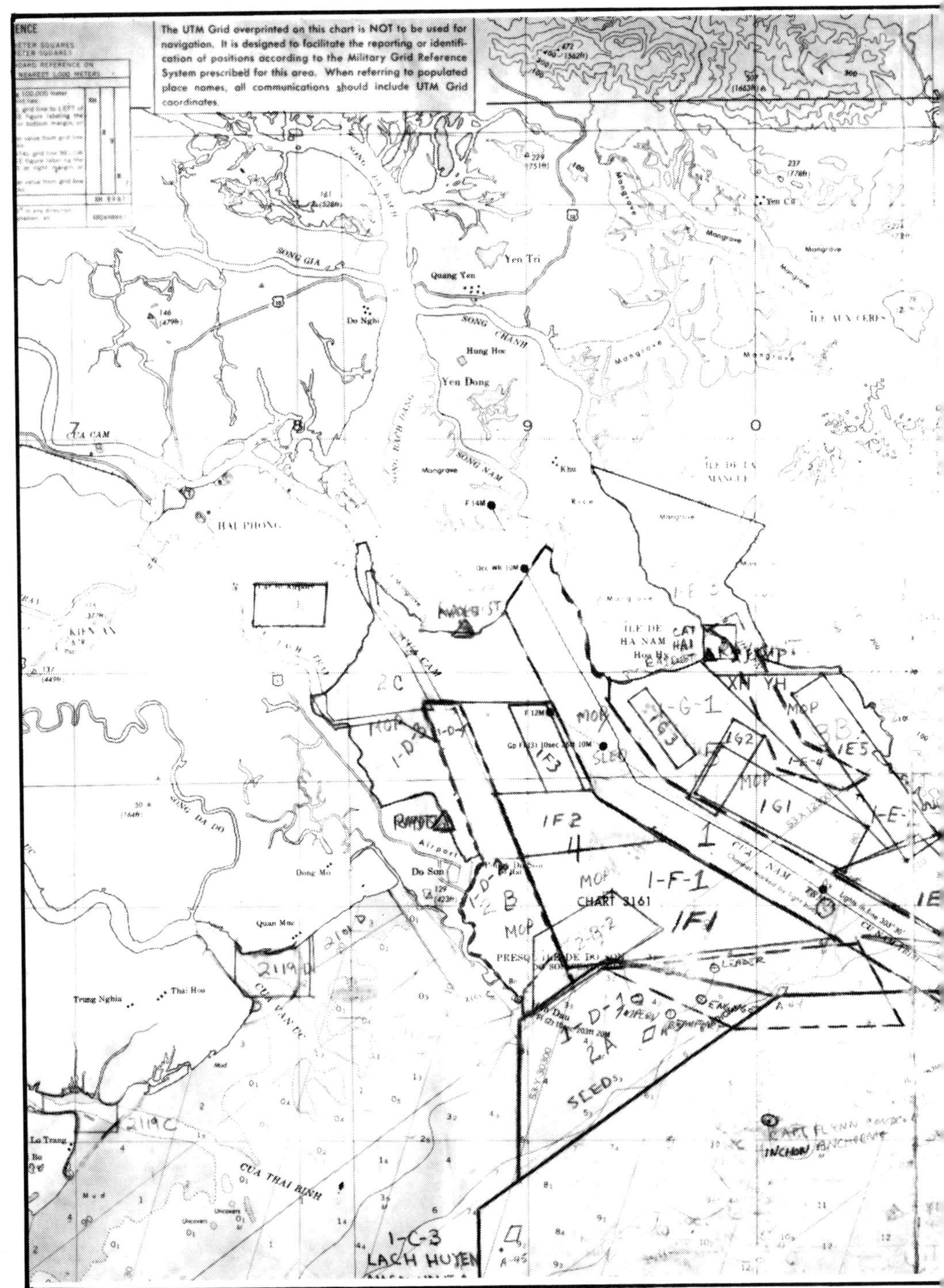

Original planning map for the clearance of Haiphong harbor.

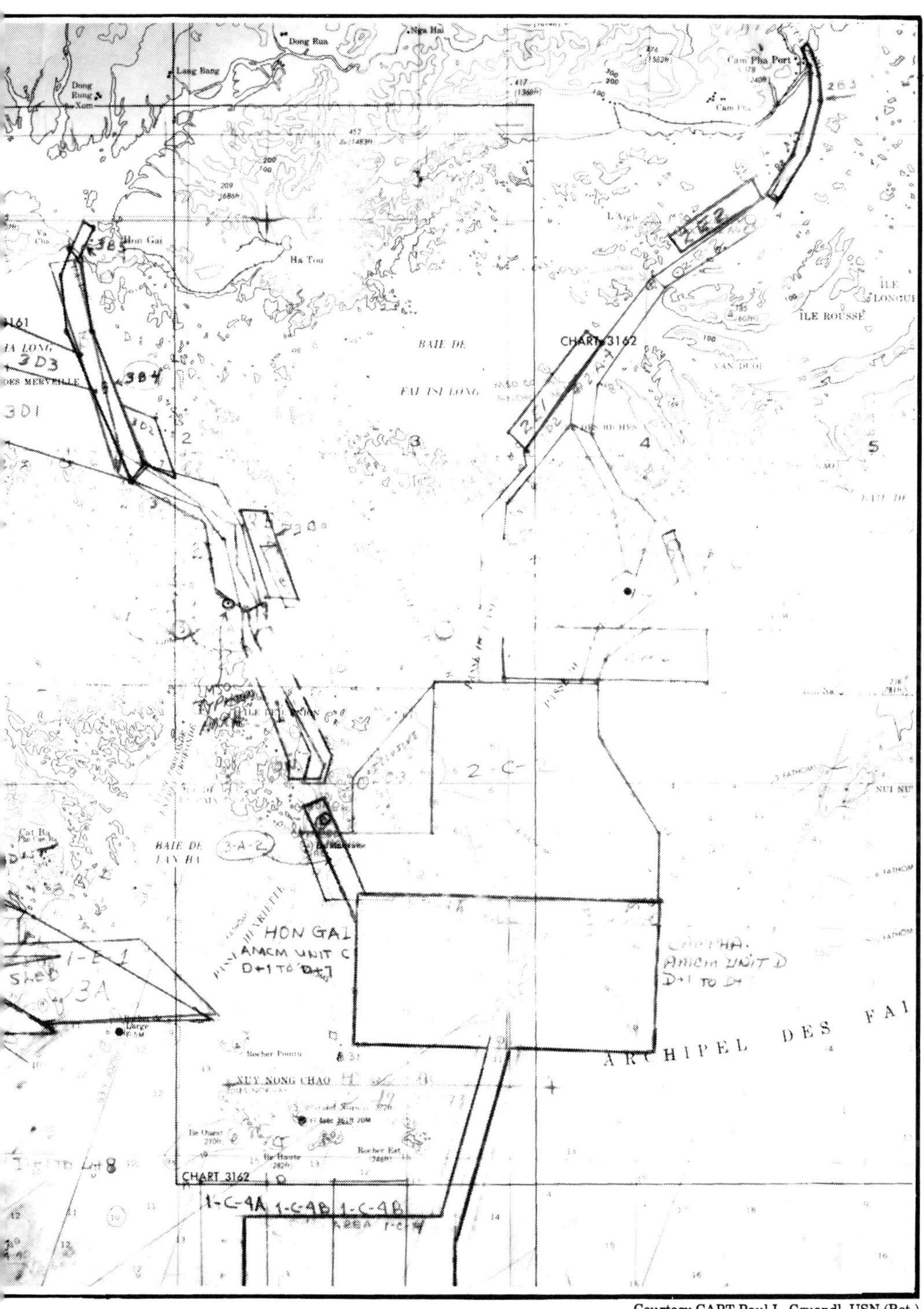

Courtesy CAPT Paul L. Gruendl, USN (Ret.)

Commander Daniel G. Powell, an exceptional surface MCM officer, and Commander Cyrus R. Christensen, a two-tour veteran of Vietnam riverine MCM and a legend in the mine warfare community for his ability to jury-rig equipment. Commander Paul L. Gruendl, one of the original planners of the mining and clearance campaign, became task force chief of staff. With half of the MINEWARFOR organization slated to embark as his operational staff, Admiral McCauley commanded an operation that included most of the Navy's best MCM officers, technology, and equipment.[27]

In formulating the clearance operation that came to be known as End Sweep, planners put highest priority on the safety of personnel and equipment. To avoid exposing the few qualified pilots and scarce equipment to live mines, the task force swept where mines had already gone inert in intentionally planned sequential tracks. Two sweep methods were planned. In an area where mines were known to have passed their sterilization dates, a check sweep consisting of a few passes with the minesweeping gear was designated. For a mined area with an unknown active status, a more thorough

U.S. Navy Photograph

Rear Admiral Brian McCauley (*front, center*), Commander Task Force 78, and Commander Paul L. Gruendl (*foreground*), his Chief of Staff, with other members of the Mine Countermeasures Force, U.S. Seventh Fleet.

Captain Felix S. "Hap" Vecchione, Commander Mobile Mine Countermeasures Command and operational Commander Task Group 78.1, Operation End Sweep.

U.S. Navy Photograph

clearance sweep was planned. In addition, the task force employed an MSS as a demonstration ship to check sweep the main channel. Ten MSOs swept deep-water approaches and acted as helicopter control ships.

With MCM for once a top priority in the Department of Defense, MCM commanders wasted no time in buying improved navigational systems and other off-the-shelf technology that they had always needed but could never get. For example, the task force bought the Raydist navigation system, a commercial electronic system that relied on the availability of a friendly land-based station to accurately vector helicopters for minefield clearance.[28]

Admiral Isaac C. Kidd, Chief of Naval Material, established a Mine Warfare Project Office (PM–19) to control scientific input into End Sweep from the various laboratories and engineering and systems commands involved in MCM. Drawing on the research resources at Panama City; Dahlgren, Virginia; and the Naval Ordnance Laboratory, the task force benefitted from the development of shallow-water sweep gear, precision navigation equipment, and integration of Raydist with the Swept Mine Locator systems. The task force could now pinpoint the exact location of mines cleared, allowing the minesweepers to discover the configuration of the minefield and to plot and to correct their sweep tracks daily.[29]

USN 1155763

A CH–53D Sea Stallion helicopter lifts a Magnetic Orange Pipe off the amphibious assault ship *Inchon* (LPH–12) in Subic Bay, Philippines, during preparatory exercises for Operation End Sweep.

The two regular detachments of HM–12 were supplemented by two additional detachments of Marines from Marine Heavy Helicopter (HMH) Squadron 463 and Marine Medium Helicopter (HMM) Squadron 164; HMM–165 provided logistic support. To prepare the pilots for the Haiphong sweep, Captain Vecchione set up an exercise area off Charleston. Buoying off the planned sweep areas in a channel with underwater approaches nearly identical to those at Haiphong, the fledgling AMCM units were able to plan and practice the sweep of Haiphong off Charleston months in advance of the actual event. After the exercise one major change was made in the clearance plans. Because of the inexperience of the borrowed Marine Corps pilots in towing the heavy Mk 105 hydrofoil sleds and the critical shortage of this expensive equipment, Vecchione pushed for development of a lighter magnetic device to be used for precursor sweeps. A scientist from Panama City devised the Magnetic Orange Pipe (MOP), a buoyant orange, styrofoam-filled, magnetized pipe that was an updated version of a World War II iron rail sweep easily towed by any pilot.[30]

USN 1155300

Task Force 78 ships in the Gulf of Tonkin head for North Vietnamese waters to begin Operation End Sweep.

On 24 November 1972, Task Force 78 was officially activated under Admiral McCauley. While the Paris peace talks proceeded, AMCM units began secretly moving west. As HM–12, MSOs, Marine helicopter squadrons, MOMCOM, Task Force 78 staff, and an armada of support vessels gathered in the Philippines, the Paris Peace Talks broke down, and aircraft of the Seventh Fleet reseeded Haiphong's minefields. The Navy and Marine pilots practiced towing the new minesweeping equipment in Subic Bay, Philippines, over the holidays and waited for orders to begin the sweep.

U.S. Navy Photograph

An automatic mine locator camera on board a minesweeping helicopter took this photograph, believed to be the only known mine swept at Haiphong.

In January 1973 a cease-fire was finally negotiated in Paris based on the exchange of all American prisoners of war for U.S. withdrawal from South Vietnam and clearance of the mines laid in North Vietnamese waters. In the Protocol signed in Paris by Le Duc Tho and Henry Kissinger on 27 January 1973 the United States agreed to meet its treaty obligation by "rendering the mines harmless through removal, permanent deactivation, or destruction." Task Force 78 left Subic Bay for Haiphong on the following day.[31]

On 27 February HM–12 executive officer Commander Jerry Hatcher flew the first AMCM mission in Haiphong's main shipping channel. The following day Nixon suspended MCM operations because of North Vietnamese delays in releasing American POWs. Although the task force resumed clearance sweeping of the northern ports on 6 March, no mine detonations occurred during the first three days. After considerable comment by both the American and North Vietnamese press about the lack of detonations, one mine did explode on 9 March and was recorded on film by the helicopter's Swept Mine Locator. On 17 March the task force swept the ports of Hon Gai and Cam Pha. *MSS–2*, formerly *Washtenaw County*, pumped full of polyurethane foam and padded for protection of the six-man, topside-only crew, ran the Haiphong channel to assure clearance. Before completing the check runs, however, MCM forces were withdrawn in protest over violations of the cease-fire in Laos and Cambodia by North Vietnamese forces. On 17 April Task Force 78 moved out to sea and then went into upkeep at Subic Bay.[32]

On 13 June both parties signed the Paris Joint Communique requiring the United States to resume minesweeping within the week and to complete clearance within the month. As all mines were past their longest sterilization date, U.S. negotiators had no difficulty promising completion of the sweep by mid-July. By 20 June the task force finished the check sweep of the main channel at Haiphong. After completion of the negotiated number of sweeps in all ports, Admiral McCauley notified the North Vietnamese that U.S. forces had "concluded" MCM operations, six months to the day after clearance had begun. Total cost of the MCM operation, including two helicopters lost, and all materials, maintenance, and casualties, was nearly $21 million, more than double the cost of the mine planting.[33]

As the task force stood down, members of the staff studied the End Sweep operation thoroughly, producing lessons-learned documents, writing short studies, and holding a symposium to reassess the operation. They concluded that End Sweep itself was not a definitive test of the new AMCM technology, as too few mines had remained active. Helicopters did sweep three to six times faster than the MSOs but suffered high equipment stress, long downtime, lack of dedicated support ships, and difficult supply logistics. What the operation did prove was that AMCM vehicles could not clear mined waters without a complex support system of surface MCM vessels, amphibious mother ships, and a strong logistic chain. Air assets were as labor intensive as surface ships because AMCM required so much support.[34]

End Sweep benefitted from circumstances not usually found in combat minesweeping operations. It enjoyed high political visibility, exceptional staff work, a large amount of lead time, and sufficient preparation by the planners. Coordination of planning, readiness, and operations with all commands at every level made it a textbook operation in effective mine clearance. Able to preempt every air-capable amphibious vessel in the Seventh Fleet and to employ every known MCM asset, the operation also benefitted from strong support of the fleet commander, shore support facilities, and a community of exceptional officers.[35] The operation was in every way a resounding success, and an unusual situation. As Admiral McCauley stressed,

> It would be a mistake to attempt to devise general, long-standing mine warfare conclusions from the specific operational and political arena in which End Sweep was conducted. End Sweep was a unique solution to a unique problem and did not present a challenge of nearly the magnitude that can be expected in the future. The location, type and settings of all mines were known. The vast majority of mines were the DST–36, a very sensitive magnetic or acoustic fuze placed on a 500 pound aircraft bomb. The magnetized pipe (MOP) was effective against this mine. It will not counter properly designed sophisticated mines... The objective of the sweeping was largely accomplished prior to laying the mines when the self-destruct time was set into the fuze.[36]

More important, End Sweep proved yet again that minesweeping, either by aircraft or by surface ship, was not by itself the answer to the problem of countering mines. "Perhaps the most important lesson learned in End Sweep," remarked Admiral McCauley, "was one that we must continually relearn. Mine sweeping of any sort is difficult, tedious, lengthy, and totally devoid of glamor."[37] McCauley cautioned that air assets failed to counter deep-water mines and were unusable in night sweeping, whereas the aged MSOs proved highly capable at both. He recommended that a balanced surface-air fleet and a varied MCM research program be pursued, particularly for development of remote-controlled guinea pig sweeps, minehunting equipment, buried mine locators, environmental data collection techniques for predicting possible future mined areas, and improved surface and air MCM assets.[38] "There remains a firm need for a balance of air and surface MCM forces," the admiral said. "This, perhaps, is the greatest lesson to be learned."[39]

The successful outcome of End Sweep, however, gave many Americans the impression that AMCM units, due to their effectiveness and mobility, had replaced surface ships as the future mine countermeasures platform. In reality the sweep had required employment of all U.S. Navy air MCM assets and conversion of twenty-four Marine helicopters for six months to check sweep two known types of mines. The major lesson learned, the effectiveness of using mines to close ports, should have triggered renewed focus on MCM requirements, but none of Admiral Zumwalt's "unions" pressed for mine force interests. The mining of Haiphong and Operation End Sweep, however, had captured the attention of the American public in a way that the more dangerous day-to-day MCM operations in the rivers of Vietnam never had.[40] As Admiral Kidd complained when End Sweep operations began, "Minesweeping seems to acquire sex appeal once every 25 years. The intervening hiatus is quite a hurdle to overcome."[41]

AMCM would remain "sexy," at least for the short term. HM–12 had just received its first shipment of new RH–53D minesweeping helicopters and began integrating them into the squadron when the Arab-Israeli War opened in October 1973. When the war ended, wreckage mixed with mines and unexploded ordnance that had closed the Suez Canal for nearly six years remained to be cleared. As part of an international agreement to clear the canal, Commander Sixth Fleet established Task Force 65 (Mine Countermeasures Force). Admiral McCauley, still serving as COMINEWARFOR, was named Commander Task Force 65 and assigned command over the international effort. Composed of members of all the services, the task force cleared land and water approaches of hulks and ordnance in a series of operations: Nimbus Star (minesweeping), Nimbus Moon Land (shore EOD), Nimbus Moon Water (underwater EOD), and Nimrod Spar (salvage).

Plans for Nimbus Star were quickly conceived, but no intelligence was available concerning the numbers or types of mines, if any, planted in the canal.[42] In spring 1974, HM–12 and MOMCOM, operating from amphibious assault ships *Iwo Jima* (LPH–2) and *Inchon* (LPH–12), swept 120 square miles of water in 7,500 linear miles of sweep track from Port Said to Port Suez with the Mk 105 sleds in a little over one month. No mines were detonated in this sector. Costs of the operation reached $4.6 million. Elsewhere, in Operation Nimbus Moon Water, a joint team of American, British, French, and Egyptian EOD specialists, under the direction of Admiral McCauley and his relief, Rear Admiral Kent J. Carroll, cleared 8,500 pieces of underwater ordnance amounting to sixty tons in eight months, including shells, bombs, grenades, and mines dating back to World War II.[43]

It was possible for the Navy to learn the wrong lessons from End Sweep and Nimbus Star. In the first three years of operation COMINEWARFOR, MOMCOM, and HM–12 completed two major mine clearance operations, operated off the decks of nearly all the Navy's amphibious ships, and served under three unified commanders, three Navy component commanders, and all four numbered fleet commanders, winning unit citations and international praise—while sweeping only one mine.

Focusing on the quick and relatively casualty-free clearance of Haiphong and the Suez Canal, the Navy declared the helicopter minesweeping a

USN 1164331

A magnetic hydrofoil sled, towed by a Sea Stallion helicopter of HM–12 during sweep operations in the Gulf of Suez.

resounding success. Admiral Zumwalt claimed that "the ability of the helicopters to sweep areas much faster than surface ships and with less manpower demonstrated that this concept was a winner." In actual practice AMCM required much more personnel, and HM–12's operational requirements resulted in the establishment of two additional squadrons for other concurrent deployment, HM–14 and HM–16.

The Vietnam War and Admiral Zumwalt's policy decisions were the death knell of the MCM surface fleet, and the number of ships on active service declined from sixty-four in 1970 to nine in 1974. Reduction of the surface MCM force had deeper ramifications for the MCM and surface ship communities. From the height of the Wonsan building program to the mid-1970s, nearly one hundred lieutenant command at sea billets disappeared with the surface MCM ships. Loss of these command billets destroyed upward mobility and a sustainable career path for officers within the MCM specialty. The consequences were a reduced presence of surface warfare officers in the mine force and lower mine consciousness in the Navy as a whole.

Still, organizationally the MCM force was in the best shape it had ever been. In addition to the establishment of MINEWARFOR, AMCM squadrons, and MOMCOM and more interest in mine warfare at the OPNAV level, the creation of the Naval Sea Systems Command in 1974 consolidated most technical mine warfare matters previously divided among the bureaus. Only MCM ship acquisition and combat systems development remained in separate offices.[44] Thus even while MCM surface forces and billets declined in real numbers throughout the early 1970s, MCM developed into an integrated, multiplatform warfare specialty, primarily because of the capabilities of MINEWARFOR. Zumwalt's greatest achievement in mine warfare was not the establishment of the primacy of AMCM; it was centralization of MCM command in the creation of MINEWARFOR.

That too soon ended. In the presence of declining budgets and higher priorities, the new CNO, Admiral Holloway, consolidated type commands in both fleets. All ships formerly assigned to the cruiser-destroyer, amphibious, service, and MCM type commands were transferred to Commander Naval Surface Force, U.S. Atlantic Fleet (COMNAVSURFLANT), or Commander Naval Surface Force, U.S. Pacific Fleet (COMNAVSURFPAC). Helicopters remained under the command of naval air force type commands in the Atlantic and Pacific fleets. Inevitably the new surface commands came to be dominated by cruiser-destroyer officers for whom mine warfare and MCM were comparatively low priorities.[45]

In place of MINEWARFOR, Admiral Holloway established Mine Warfare Command (MINEWARCOM), ultimately a one-star command at Charleston. Strictly defined as a technical advisor to the CNO and liaison to the operational fleet commanders, Commander MINEWARCOM advises the

CNO on all mine warfare matters including readiness, training, tactics, and doctrine and coordinates MCM matters with the fleet commanders controlling air, surface, and undersea mining and MCM assets. COMINEWARCOM also provides mine warfare training for the combined Fleet and Mine Warfare Training Center and acts as mine warfare liaison to individual laboratories that report their mine and MCM advances directly to the Chief of Naval Research. The only operational units under the direct control of COMINEWARCOM are the Mobile Mine Assembly Groups (MOMAG), which are responsible for storing, maintaining, and assembling mines through thirteen active and twenty-seven reserve mine assembly detachments worldwide, and the Mine Warfare Inspection Group, which regularly trains and inspects 350 Navy units in the tactical and technical aspects of minelaying and MCM. In addition to these duties, COMINEWARCOM often commands Charleston Naval Base.[46]

The cumulative effect of the organizational and policy changes implemented under Admirals Zumwalt and Holloway was a de-emphasis on MCM within the Navy. For four brief years, 1971–1975, mine warfare had a coordinated command structure that allowed an increasingly complex service to deploy quickly and use effectively all the diverse elements required to counter mines. After 1975 many of the requirements for a comprehensive MCM capability were gone. Only a handful of MCM ships remained in active service. The critical on-scene coordination of MOMCOM to conduct AMCM operations no longer existed, and the training and readiness of the MCM force were unequal competitors in newly formed type commands. Henceforth, the Navy was once again committed to improvisation when faced with the need for serious MCM.

As the U.S. Navy's surface, air, and submarine communities continued to develop and to integrate their warfighting skills to meet the anticipated Soviet threat, the Soviets advanced their own interest in mine warfare, developing new influence mine actuating systems. Continuous microprocessor enhancements improved mine selectivity, allowing mines to differentiate between real and false targets and determine whether to detonate, hesitate, or abort on an individual basis. Command-detonated mines were developed with remote controls, and deceptive stealth mines, designed to naturally blend into the underwater environment, soon made some mines almost impossible to detect visually in shallow water.

American mine experts developed their own "smart" mines. For deep-water mining, American scientists created the Captor (Encapsulated Torpedo) mine, a moored mine armed with a modified homing torpedo. Using both passive and active sonar to locate targets, Captor mines remain dormant until activated by a passing ship. Such increasingly sophisticated mine technology forced MCM experts to search for a way to counter such mines, which can lurk and attack specific targets.[47]

With the development of smart and rising mines, American strategists began to realize that Soviet mines were not merely a threat limited to relatively shallow waters. In the 1960s the Soviets had demonstrated their growing capability to lay mines in previously unmineable shipping lanes and choke points to counter U.S. Navy submarine and carrier operations. The threat of deep-water Soviet mines revitalized limited interest in surface MCM for open-ocean operations, but this interest sputtered along for several years without coming to fruition. Plans for an improved, wooden hull MSO–523-class were shelved in the late 1960s due to continuing budget constraints; not until the mid-1970s were ocean minesweepers reconsidered.

In 1976 Admiral Holloway approved the design for a steel-hulled, deep-ocean minesweeper-hunter to supplement AMCM helicopters' shallow-water capability. Vice Admiral James H. Doyle, DCNO for Surface Warfare, recommended a limited shipbuilding program for this specific capability. The Navy proposed building nineteen of these steel-hull ships, but the Carter administration delayed the program, and the ship was never built. Planners then recommended a low mix of smaller, faster vessels, particularly small MSBs and MSLs that could be carried by an amphibious mother ship.[48]

A compromise package developed in 1979 under the personal leadership of CNO Admiral Thomas B. Hayward called for integrated minehunting and clearance systems on a number of different platforms at much lower cost and size. The systems would be centered around a deep-ocean mine countermeasures ship (MCM) as a replacement for the MSOs. In addition to the MCM with its modern autonomous mine neutralization system (MNS)—an advanced remotely operated minehunting vehicle—a new, smaller class of coastal minesweeper hunters (MSH), with most of the minehunting capability of the MCM, was proposed.

While the Navy and the Congress debated the size of this theoretical MCM building program, Admiral Hayward tasked COMINEWARCOM with the development of a Craft of Opportunity Program (COOP) to employ reserves and minesweepers. Using rented or confiscated shrimp boats (former drug runners) with the proper configuration for deep-water trawling, MCM veteran Captain Cyrus Christensen directed Mine Squadron 12 to experiment in re-equipping the vessels with more powerful generators required to operate the influence minesweeping gear taken from scrapped MSOs and mechanical sweep equipment formerly used on MSCs.

Informally rechristening their first effort the minesweeping shrimp boat (MSSB–1), Christensen tested commercially available sonar, navigation, radio, and minehunting equipment at Charleston. Encouraged by the tests, he obtained assistance from the trawler conversion experts of the Royal Navy to develop deep trawl equipment and proved the viability of fishing boats in channel surveying, minehunting, and sweeping.[49] Determined to prove to the Navy that drastically reduced resources were decreasing the combat readiness of the MCM forces, Christensen obtained experimental models of several types of advanced Soviet mines for demonstration. In breakout exercises in 1979 and 1980 Christensen proved, to the consternation of many senior officers, that "all of the forces the U.S. Atlantic Fleet could bring to bear could not open one East Coast port in any acceptable period of time." [50]

In 1980 COMINEWARCOM Rear Admiral Charles F. Horne III brought the MCM community's program for active recognition of the Soviet mine threat to the halls of the Pentagon. Admiral Horne succeeded in convincing many key players in the Pentagon of the seriousness of the Soviet and Third World mine threat and the inadequacy of the remaining U.S. surface assets. Appreciating that sophisticated MCM required a mix of versatile platforms, technology, and personnel, Admiral Hayward personally encouraged an across-the-board MCM "renaissance." Such a program would include advances in computer-assisted threat evaluation, training, advanced sonar and mine neutralization vehicles, new ship construction, refits of older MSOs, and new, larger, faster, and more adaptable AMCM vehicles that could tow heavier equipment, and operate for longer periods and at night. In addition, Hayward approved COMINEWARCOM's proposal that a reserve harbor defense program be developed along the lines of the existing British trawler force and the MSSB–1 experiments. Admiral Hayward tasked COMINEWARCOM to develop the COOP program, and integrate it into the buildup of the 600-ship Navy planned by the Reagan administration. Plans originally called for twenty-two units, one for each key U.S. port, with four reserve crews assigned per unit. [51]

While the post-Wonsan Navy had developed some technological advances in minehunting techniques, particularly mine-classification variable-depth sonar, by the mid-1960s the Navy had fallen far behind its NATO allies in MCM development, particularly in fiberglass shipbuilding technology and small drone boat minesweeping systems. European nations, notably France, had successfully developed increasingly sophisticated versions of tethered dual minehunting and mine-neutralizing ROVs. On the other hand, U.S. industry had developed many excellent potential ROVs for the offshore oil industry. The U.S. MNS, planned for the new MCM ship, would be built upon these foreign and industrial advances, employing both sonar and television cameras to locate, classify, and neutralize mines with explosives or cable cutters. [52]

Committing itself to a 31-ship, MCM building program of both high-performance, high-cost and low-performance, low-cost vessels, in 1981 the Navy proposed a sophisticated fiberglass-encased wood-laminate design, the 224-foot *Avenger*-class MCM, to support both deep-water mechanical minesweeping and the advanced minehunting capabilities of the developmental MNS. In addition to having mechanical and influence sweep gear, the *Avenger* class is equipped with advanced minehunting sonar, precise

Courtesy Peterson Builders, Inc.

Avenger, the first ship of the MCM class, was designed primarily as a deep-water minesweeper-hunter.

integrated navigation system (PINS), and high-definition surface-search radar—all integrated to create a flexible platform from which a variety of minehunting tasks can be developed. These *Avenger*-class ships, which have recently entered the fleet, are currently considered the most advanced MCM vessels in the world.[53] On the low-cost end, the *Cardinal*-class MSH, a small fiberglass surface effect ship using air-cushion technology, was designed both to hunt and to sweep mines in coastal waters.

Thus in 1981 the Navy committed to a new shipbuilding program that would revitalize surface MCM forces with modern, high-technology ships and MCM systems. Unfortunately so much time had passed since the last MCM ships had been built that the techniques of wooden and MCM shipbuilding had been forgotten and had to be relearned. Technological advances in fiberglass construction, sensors, and sonar could, however, be borrowed directly from allied nations and U.S. industry. On the high-cost end the first *Avenger*-class MCM ships ran into program delays caused by well-meaning cost-cutting attempts to use stockpiled main propulsion engines from previous programs and over 17,000 design alterations on the first versions alone. On the low-cost end, MSH hull sections delaminated during testing, and the Navy terminated development of the *Cardinal*-class design in 1986, adapting instead the design for an existing Italian MCM ship, *Lerici*, as the MHC–51 *Osprey*-class coastal minehunter.[54]

Regular delays associated with the MCM program and cancellation of the MSH reminded some elements of the Navy of the hazards of a quick-fix approach to developing MCM technology. In the long time between building programs, much of the art had been lost, and regaining it proved costly. As one shipbuilding expert noted,

> The principal cause of our current difficulties in mine warfare shipbuilding is the lack of long term, sustained production program through the 1950s, 1960s and 1970s. We are now reinventing modern shipboard mine countermeasures after a 30 year gap instead of having it continuously evolved at a measured pace.[55]

As the rebuilding program moved slowly forward, the leaders of the MCM community worked to maintain the readiness of their force and its proper place in the warfighting consciousness of the Navy. They were often frustrated by the glacial progress of both. AMCM units continued to train for quick deployment and experimented with towed minehunting sonar, but did so without a dedicated support ship; AMCM squadron commanders had difficulty even obtaining deck time to train their men. MCM priority was no higher within DOD. Navy RH–53D minesweeping helicopters were chosen for the April 1980 Iranian hostage rescue mission, although minesweeping had nothing to do with the mission. When seven of the Navy's inventory of thirty minesweeping helicopters were lost, Secretary of Defense Harold Brown hesitated to replace them, reinforcing the impression that few in top authority thought MCM of national importance.[56] Successive MINEWARCOM commanders sought to invigorate the Navy's interest in all facets of MCM but succeeded in little more than sustaining the status quo in numbers of future ships to be built. That a fully integrated MCM force was indeed a necessary part of warfighting was a lesson the Navy would have to learn and relearn.[57]

Ten years after the clearance of the Suez Canal, mining activity in the Red Sea and the Persian Gulf again brought MCM into international prominence. In July and August 1984 suspicious underwater explosions crippled at least sixteen merchant vessels in the Gulf of Suez. In response to Egyptian appeals, advisors from U.S. Central Command (CENTCOM), EOD personnel from Commander in Chief, U.S. Atlantic Fleet (CINCLANTFLT), and MINEWARCOM staff joined an international mine hunt to search for the source of the explosions. British minehunters already in the area were assigned to a sector requiring intense minehunting capabilities, while the Italians and the French, whose forces included former American MSOs converted to minehunters, took smaller search sectors. Egypt, seeking a complete mix of forces, specifically requested U.S. AMCM helicopters, and HM–12 and HM–14 immediately stood by on alert. Within hours of the official request for assistance the first of the helicopters were en route on U.S. Air

Force C–5s. U.S. forces dubbed their duty Operation Intense Look.[58]

As the mine crisis arose during the annual Moslem pilgrimage to Mecca, Saudi Arabia requested emergency assistance in sweeping the ports of Yanbu and Jidda, and the U.S. forces split into two detachments. The first detachment, supported by Middle East Force flagship *La Salle* (LPD–3), swept these ports and also swept the Bab el Mandeb with combined magnetic and acoustic hydrofoil sleds to ensure safe passage for carrier *America* (CV–66) and her escorts transiting from the Indian Ocean to the Mediterranean.

A second detachment designated to sweep in the Red Sea was assisted by the coastal hydrographic survey ship *Harkness* (T-AGS–32). It was supported by an assigned Atlantic Fleet EOD side-scan sonar detachment and the amphibious transport *Shreveport* (LPD–12) as a helicopter platform. Helicopters from the second detachment towed the new AN/AQS–14 minehunting sonar. In this first operational deployment of the AQS–14 the detachment flew up to eight missions in their assigned sector of the Red Sea despite heavy weather. As an MCM operation, the effort was a success, for repeated sweeps proved the areas free of mines. As an operational test of the new sonar, however, the results were inconclusive; no mines were located by the U.S. MCM forces in these waters.[59]

Several mines were detonated by the international forces, but the British recovered and exploited one advanced combination influence mine based on a Soviet design and believed to have been laid by the Libyans. Although the clearance operation underscored the growing international fears of undersea terrorism, the terrorist threat was doubly apparent to the men of HM–14 whose homecoming was delayed by emergency support and medical evacuation missions to the bombed U.S. Embassy in Beirut.[60]

In light of the ease with which terrorists demonstrated their ability to mine this important international choke point, MCM quickly became the focus of international concern. Studies soon noted the importance of coordination of international MCM forces and national integration of mobile air, sea, and undersea MCM forces, the lessons repeatedly learned by U.S. MCM forces since Wonsan.[61] The overall effect of such low-intensity mine warfare by terrorist organizations and the Third World reminded many nations of their own vulnerability to mines.

Escalation of such low-intensity mining in Middle Eastern waters continued to impinge on U.S. interests throughout the 1980s. From the beginning of the Iran-Iraq war in 1980 merchant vessels suffered regular air, surface, and mine attacks by both sides in the transit lanes in the Persian Gulf. Escalation of attacks on commercial vessels in 1984 led Kuwait to request convoy protection from other nations for its tankers. When the Soviets agreed to assist Kuwait in 1986, the Reagan administration reconsidered and decided to offer protection to half of Kuwait's tanker fleet by sailing them under the American flag with appropriate military protection. While Congress openly debated the decision to reflag Kuwaiti tankers, new fields of moored contact mines began appearing throughout the gulf. These were, by and large, M–08 mines, manufactured in North Korea of 1908 Russian design and laid by the Iranian Revolutionary Guard. Soon, these mines began to break free of their mooring cables and joined similar MYaM contact mines, laid by both Iran and Iraq in the early 1980s, littering the gulf waters and threatening U.S. warships.

In early 1987 U.S. Navy EOD divers, supported by Kuwaiti and Saudi Arabian surface forces and by other U.S. Navy EOD personnel mine spotting from a Kuwait Air Force helicopter, cleared ten contact mines near Kuwait's Al-Ahmadi oil terminal. COMINEWARCOM immediately deployed an evaluation team to Kuwait while COMNAVAIRLANT put HM–14 on 24-hour alert.[62]

The first convoy of reflagged tankers protected by U.S. Navy warships under the codename Operation Earnest Will directed by Commander Middle East Force (CMEF) began steaming toward Kuwait on 24 July 1987. Supplied with the exact route and timing of the first convoy, Iranian Revolutionary

Courtesy RADM W. W. Mathis, USN

SS *Bridgeton*, 24 July 1987, four minutes after a mine blew a 35-by-45-foot hole in her port bow. She shows a list of only about one degree and lost no speed capability as a result of the damage.

120

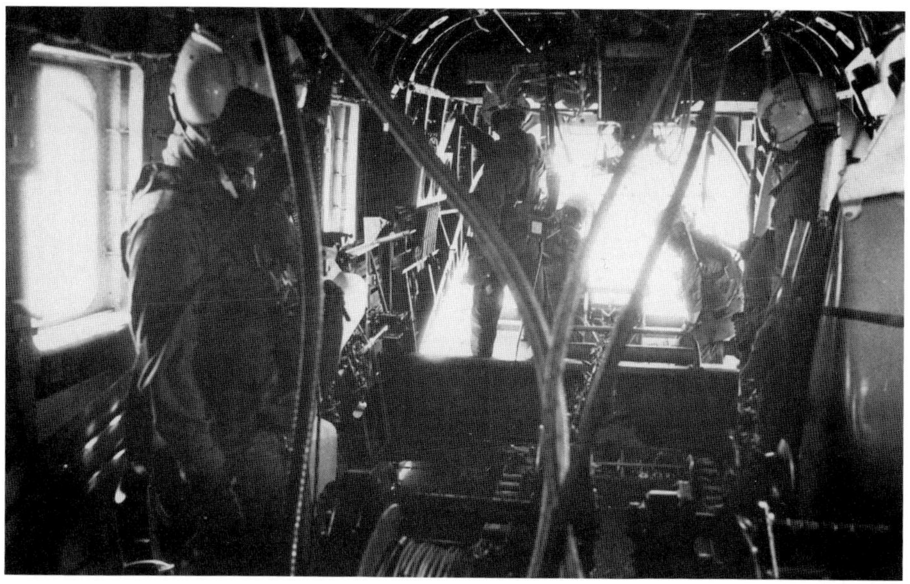

Interior view of an RH–53D helicopter during Persian Gulf minesweeping operations, August 1987.

Guards found mining the supertankers' path an easy matter. As escort ships steamed in front and astern, SS *Bridgeton* hit an M–08 mine off Farsi Island, damaging but not stopping her. Escorting cruiser *Fox* (CG–33), frigate *Crommelin* (FFG–37), and destroyer *Kidd* (DD–993) fell in behind the supertanker, and the convoy continued on to Kuwait. Americans watching nightly television newscasts saw the U.S. Navy escort vessels in a scene reminiscent of Farragut's ships at Mobile Bay, steaming in column behind *Bridgeton* and SS *Gas Princess* to take advantage of their deeper draft as a precaution against additional moored mines. For *Bridgeton*'s return trip U.S. MCM advisors outfitted two 150-foot Kuwaiti commercial tugs, *Hunter* and *Striker*, with standard MSB mechanical minesweeping gear. When one-third of the tugs' regular civilian crews refused to undertake minesweeping, the Navy recruited volunteers to man the minesweeping tugs from among the experienced enlisted men serving ashore at the Administrative Support Unit, Bahrain, and from U.S. MSB units.[63]

HM–14, which had been on 24-hour alert status in Norfolk for over a month, received word to uncrate and unpallet their equipment and to return to their standard 72-hour alert just days before *Bridgeton*'s mining. No sooner had they done so, than they were called to the gulf; in six hours their first load of equipment headed east.[64] When HM–14's detachment of eight RH–53D helicopters arrived, they began minesweeping operations in advance of the convoys and were supported, in turn, by amphibious assault ships

DN-SN–88–01078

Divers head toward *Illusive* (MSO–448), one of three ocean minesweepers in tanker escort operations in the Persian Gulf, fall 1987.

Guadalcanal (LPH–7) and *Okinawa* (LPH–3). Operating in convoy for the first time, the squadron quickly realized that the surface ships had difficulty adjusting to the peculiar demands of the large minesweeping helicopters. The logistic needs of the AMCM helicopters forced them to carry many sets of duplicate gear and parts, and the decks of the ships were barely large enough to contain their equipment. Heat, humidity, and caking sand cut the helicopters' operating time on convoy, but the detachment swiftly got the aircraft back in operation and visually searched for mine lines in advance of the convoys. In addition to the aircraft, four MSBs unsuccessfully attempted to clear mines off Bahrain Bell, an area regularly used by U.S. vessels.[65]

Expecting a protracted operation requiring the best available minehunting and minesweeping capabilities, the Navy deployed six of the remaining MSOs, three from each fleet (five from the Naval Reserve Force), with rotating crews of active duty personnel. To save wear on their vintage ships, the three Atlantic Fleet ships, *Fearless* (MSO–442), *Inflict* (MSO–456), and *Illusive* (MSO–448), were towed to the gulf by *Grapple* (ARS–53) over 9,000 nautical miles. The three Pacific Fleet ships, *Esteem* (MSO–438), *Enhance* (MSO–437), and *Conquest* (MSO–488), proceeded to the gulf under conventional tow by an LST.[66]

The records compiled by the MSOs operating in the Persian Gulf from 1987 to 1990 are by any definition exceptional. On arrival in the gulf under the command of Captain Jerry B. Manley and successive MCM Group Commanders (MCMGRUCOM), the MSOs began operating, sweeping regular Q-route channels and minehunting with their SQQ–14 sonar in the convoy shipping lanes. The MSOs had limited success in identifying mine-like contacts by using industrial ROVs, notably Super Sea Rover, lightweight units with no mine neutralization capability, and relied on EOD personnel for mine destruction. Despite rumors that North Korea may have provided the Iranians with influence mines, MSOs found only the contact variety.

Within the first eighteen months of Persian Gulf minesweeping operations, the MSOs accounted for over fifty moored mines, cleared three major minefields, and check swept convoy tracks throughout the gulf. In March 1989, almost one year after finding their last moored contact mine, half of the MSOs returned to the United States. After thirty years in the mine force the MSOs in the Persian Gulf had a rare opportunity to operate in the manner

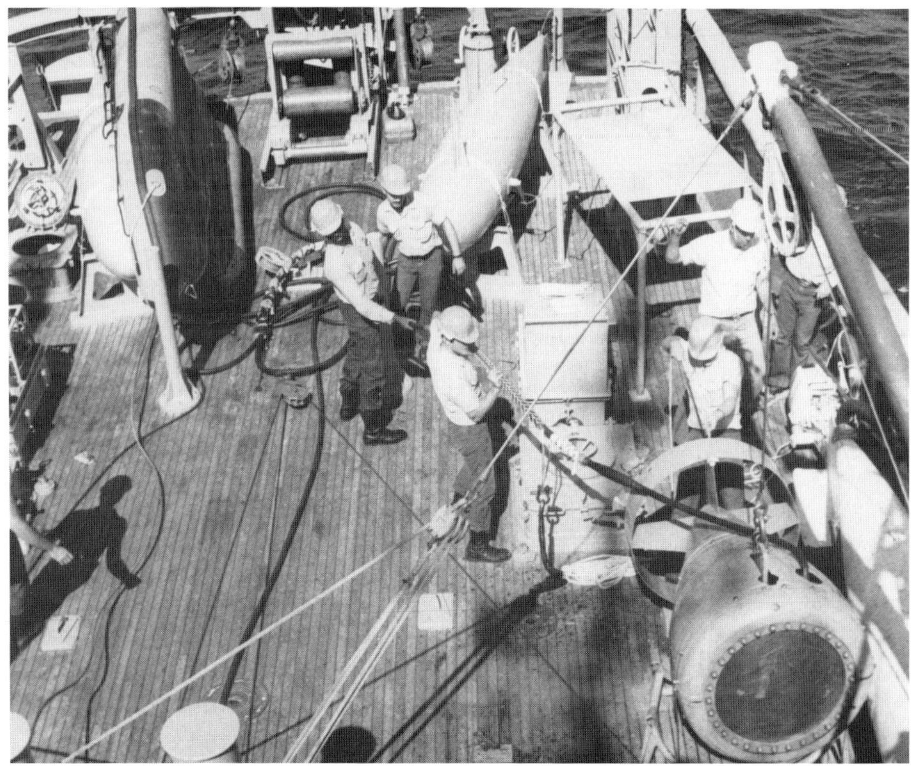

DN-SN–88–01075

Illusive's crew members get ready to lower an acoustical device during training for escort operations in the Persian Gulf, fall 1987.

for which they had been designed: to hunt and sweep mines in established routes in advance of convoy sorties.[67]

Other MCM elements operating in the gulf included advisors and scientists. COMINEWARCOM initiated a Navy Science Assistance Program, installing experimental mine-detecting devices on combat ships for protection against drifting mines and developing new minesweeping tactics to meet the threat posed by both moored and drifting mines. Several scientific MCM initiatives, such as minehunting optics and the use of underwater, unmanned vehicles for mine identification, received their first operational testing in the gulf. Fleet commanders and COMINEWARCOM provided staff officers to the MCM Group Commanders, who also advised commanding officers and Joint Task Force Middle East (JTFME) staff on mine matters, planned sweep operations, refined tactical and navigation procedures, developed maintenance and logistics schemes, and cooperated with other nations' MCM units sweeping waters nearby.[68]

The mine threat in the Persian Gulf increased U.S. Navy MCM awareness, at least briefly, and reminded captains of their ships' vulnerability to an increasing pattern of low-intensity mine warfare. "Events in the Gulf," noted one analyst, "have done much to shake navies out of their lethargy—one of the most worrying aspects of the Iranian mine warfare offensive is how

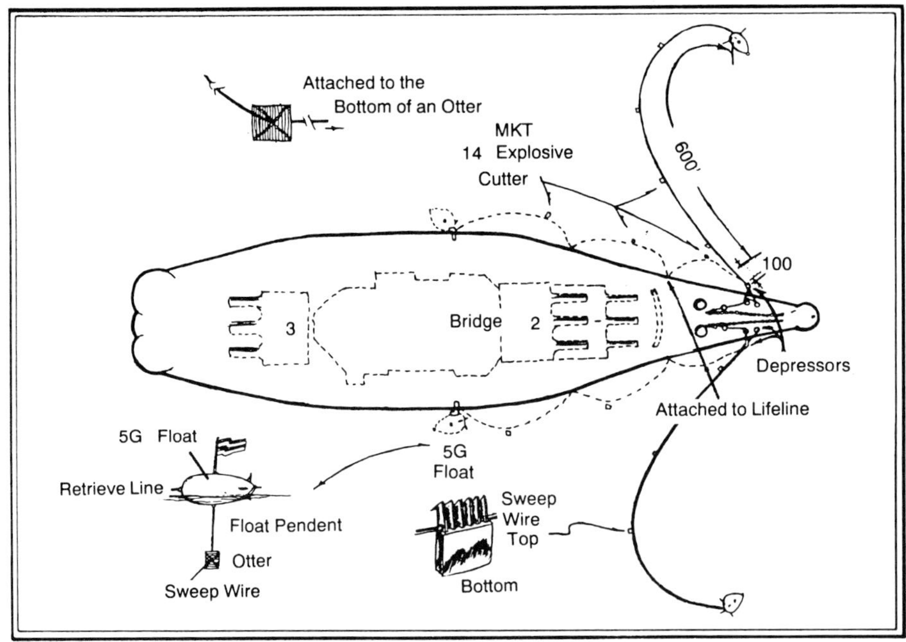

Courtesy Bobby Scott and Wilfred Patnaude

Diagram of *Iowa*'s bowsweep system as used in 1987.

limited it really is." Navy helicopters, particularly the frigates' capable LAMPS MK III SH–60Bs that were usually assigned to antisubmarine warfare missions, were regularly assigned to mine spotting in advance of ships. The standard ship protection, however, remained the bow watch, a relic of the Civil War. The temperature and salinity of the waters permitted most shallow-moored or floating mines in the gulf to be seen by a bow watch, and sailors watched carefully for mines after *Bridgeton*'s mining.[69]

For most surface officers, concern about mines was an unusual addition to regular shipboard operations. Some officers went further than others to protect their ships from mine attack. As the battleship *Iowa* (BB–61) prepared for deployment in August 1987, Captain Larry Seaquist fabricated a successor to his ship's original World War II paravane system. Aided by the COMINEWARCOM staff, Mine Squadron 2, and the Vietnam river minesweeping experience of his master chief and ship's boatswain, Seaquist rigged standard 5–G sweep gear to the bow of *Iowa* and prepared to cut mine mooring cables if necessary. When deployed in exercise, this gear proved easy to stream and the ship retained her maneuverability even at high speeds. Encouraged by Seaquist's example, at least one other officer considered streaming bowsweep equipment in the gulf to cut contact mines.[70]

In addition to providing MCM assets for clearance operations, U.S. Navy

DN-SN–88–10166

The barge *Hercules*, one of the forward-deployed mobile bases in Operation Earnest Will, had a main mission of preventing minelaying.

and joint service officers also developed an aggressive strategy to prevent mine attacks in the Persian Gulf through a combination of forward basing, continuous presence, surveillance, patrolling, and carefully measured responses to individual situations. Using the talents of special operations and staff officers who had served on mobile support bases in the rivers of Vietnam, JTFME transformed two mobile oil platforms into Mobile Sea Barges and undertook operations from the barges to prevent minelaying. Much as Vietnam had demonstrated that riverine MCM was a combined operation, the gulf war of the late 1980s proved that prevention of minelaying is an all-Navy and joint service combat operation.

Despite Iranian protests that their vessels laid no mines, U.S. intelligence assets tracked the Iranian landing craft *Iran Ajr* from its port in Iran to an area north of Qatar in September 1987. A U.S. helicopter from the frigate *Jarrett* (FFG–33), using night-vision cameras, detected minelaying activity aboard the vessel and observed the crew laying at least six mines, and fired on the ship. After a short time the ship resumed minelaying and the helicopters again fired; at daybreak a boarding party captured the boat with nine mines aboard. The publicity surrounding this event effectively halted Iranian minelaying for six months. The boarding party also recovered charts marked with minefields planted that night, allowing MCM forces to exploit the field.[71]

Iranian mines provoked a stronger measured response from the Navy the following year. On 14 April 1988, while on a routine transit between Earnest Will missions, seventy nautical miles east of Bahrain Bell, the bow watch on

DN-SC–87–12584

Contact mines on board the captured Iranian minelaying ship *Iran Ajr*.

frigate *Samuel B. Roberts* (FFG–58) spotted three shiny new mines in the water about one-half mile away. "Damn, those look just like mines," Commanding Officer Paul X. Rinn immediately thought as he ordered his ship to stop. Calling his men to general quarters and closing the ship's water-tight doors and hatches, Rinn followed the standing instructions of Commander JTFME Rear Admiral Anthony A. Less to prepare his LAMPS MK III helicopter to mark the mine position with floats and flares. While the helicopter prepared to launch, Rinn began warily backing down in the ship's wake away from the mine line using the ship's forward auxiliary power units. Hampered by rough seas, the ship swung into another submerged M–08 mine that blew a 20-foot hole in her hull, broke the keel, blew the engines off their mounts, and flooded the main engine room and other spaces. Trapped in a probable minefield after dark, Rinn refused assistance and fought the fires and flooding, stabilizing the ship until she could emerge from the minefield under her own power. Without the immediate and effective damage control efforts of her crew already in their proper stations, the mine damage would have sunk *Roberts*. After patching the ship together that night, crewmen trained searchlights on the water seeking more mines, another ship-protective method like the bow watch used by naval vessels since the mid-nineteenth century.

MSOs and assisting ships found and destroyed five mines in the waters around the site of *Samuel B. Roberts'* mining. Determining that the mines were of the same manufacture as those captured aboard *Iran Ajr* the previous September, the United States planned a measured response. In retaliation for the mining of *Roberts*, the United States launched Operation Praying Mantis, destroying two Iranian oil platforms and nearly half the Iranian Navy on 18 April 1988, ending at least for a time most Iranian mining attempts.[72]

MCM efforts in the Persian Gulf in the 1980s benefitted from unusually easy mine spotting, generally good weather, a lack of antisubmarine warfare problems, and manageability of the other threats. The lack of influence mines also kept casualties low and operations possible. Hence, the aging but able MSOs, helicopters, and EOD and special operations personnel dedicated to MCM efforts capably met the mine threat. That achievement, however, required full use of all the Navy's available MCM assets. Very shortly after those assets were finally removed and the remainder of the MSOs returned stateside, war returned to the Persian Gulf.

In August 1990 Saddam Hussein's Iraqi army invaded and captured neighboring Kuwait. United Nations sanctions and economic embargoes failed to convince Hussein to withdraw from Kuwait, and America and its U.N. allies prepared for a lengthy war on the land and in the easily mined waters of the Middle East. Immediate MCM planning began, AMCM and EOD units deployed, and *Avenger* (MCM–1) and three MSOs, *Adroit* (MSO–509), *Impervious* (MSO–449), and *Leader* (MSO–490), were immediately sealifted

A Navy EOD diver attaches a neutralization charge to a LUGM contact mine during Operation Desert Storm. Note that the diver is using the new Mk 16 underwater breathing apparatus (UBA), which features a low magnetic and acoustic signature.

to the Persian Gulf to provide MCM forces for Operation Desert Shield.[73] After U.N. forces commenced wartime Desert Storm on 16 January 1991, the operational tempo of the MCM forces rose markedly. Acting on U.S. intelligence estimates of the mine threat, U.S. and British surface MCM ships, as well as U.S. AMCM helicopters led by the Commander U.S. MCM Group Command (COMUSMCMGRUCOM), cleared a channel toward Kuwait for the advancing amphibious assault force.

On Monday, 18 February 1991, two U.S. Navy warships involved in these mine clearance operations, amphibious carrier *Tripoli* (LPH–10), the flagship of allied MCM operations, and guided-missile cruiser *Princeton* (CG–59), a new and expensive ship equipped with the state-of-the-art Aegis long-range air defense system, were mined in two separate incidents in the northern Persian Gulf. *Tripoli* hit a moored contact mine fifty miles off Kuwait. Although she suffered at least a 16-by-25-foot hole in her hull, she remained temporarily on station to support minesweeping operations in preparation for the anticipated amphibious assault.

A few hours later and ten miles away, *Princeton* detonated at least one influence mine under her keel, which lifted the ship out of the water, cracked her hull, and caused extensive damage to her midsection and one propeller. Although *Princeton*'s Tomahawk weapons and Aegis systems remained intact, she was unable to stay on station, and was towed to port for further damage assessment.[74] With almost daily increases in mine spotting and destruction by Navy surface, air, and EOD forces, U.S. and British MCM ships led U.S. battleships into a cleared fire support area under constant targeting by Iraqi silkworm missile launchers.

The mining of two important Navy warships in waters believed to be mine-free once again emphasized the U.S. Navy's recognized failure to sustain adequate combat MCM capability. At a congressional hearing on 21 February 1991, Secretary of the Navy H. Lawrence Garrett III explained that the Navy "spent more than 25 years not developing or buying new minesweepers or minehunters," while others pointed to continuing U.S. dependence on NATO allies for minesweeping technologies and assistance. The extreme danger to attacking allied forces posed by extensive Iraqi sea mining and coastal fortifications ended with the success of the allied ground war, assisted by the successful deception of the Iraqi defenders by the allied amphibious forces off

U.S. Navy Photograph

Damage done to *Tripoli* (LPH–10) by a primitive Iraqi mine in the northern Persian Gulf during Desert Storm.

Courtesy Mine Warfare Command

This bottom influence mine was towed onto the beach for exploitation during Desert Storm. Note the two lifting balloons that were used to raise the mine off the ocean floor.

Courtesy Surface Warfare magazine

A developmental version of the SQQ–32, the next generation of advanced minehunting sonar, was operationally tested by *Avenger* during Desert Storm with exceptionally good results.

Kuwait. As the 42-day war came to an end in late February, U.S. and British forces had accounted for more than 160 contact mines. Augmented by MCM forces of France, Belgium, the Netherlands, Germany, Italy, and Japan, independent allied forces cleared hundreds of contact and influence mines using captured charts and basic intelligence provided by Iraqi sources as a requirement of the cease-fire. Throughout the spring and summer of 1991, U.S. and British MCM forces swept and hunted hundreds of square miles, clearing channel routes to reopen commerce with Kuwait, and then joined other coalition MCM forces in attacking the mine lines.[75]

The escalation of the Vietnam War in the mid-1960s drained the Navy and the MCM fleet of substantial resources that had been carefully built up to sustain naval warfighting skills. As shallow-water riverine MCM became the focus of important day-to-day MCM operations throughout the late 1960s, fewer surface naval officers became involved in MCM, which necessarily took on a special operations character in the rivers of Vietnam. When MCM once again became a national priority, albeit briefly, in End Sweep, the commitment of the Navy to AMCM and the substantial redefinition of MCM operations required to support helicopter operations completely reidentified MCM as a specialty force.

Spurred on by improved Soviet deep-water mining capabilities and by belated recognition of the inadequate state of the existing mine force ships, the Navy began rebuilding the MCM fleet in the 1980s only to discover that the complex requirements of building nonmagnetic ships would stall their entry into the fleet for nearly a decade. The Persian Gulf wars in the late 1980s and early 1990s and the mining of *Samuel B. Roberts*, *Tripoli*, and *Princeton* reminded many naval officers of the lessons of the history of mine warfare, particularly the ease with which simple, antique mines can destroy the most advanced warships. The Navy also relearned that although reprisals for such mining attacks can immediately deter minelaying, only a flexible, combat-ready MCM force can clear mined waters. The Navy also rediscovered one of the major lessons repeatedly learned by the mine force since Wonsan: for the Navy to meet mine threats in an age of low-intensity warfare, the MCM force must be considered part of the naval warfighting team along with other communities in the Navy.

Current Navy plans call for a balanced MCM force of surface and air assets operating in tandem to clear home ports, choke points, sea lines of communication, and forward operating areas, in sequence, with AMCM assets specifically tasked for breakout and quick response. U.S. Navy operational MCM assets in 1989–1990 consisted of a total of 20 Korean War-era MSOs in both the active and the reserve fleet, 21 RH–53D helicopters, and 7 MSBs. The first new MH–53 helicopters and MCM vessels funded in the 1980s are also now entering fleet operations. New construction plans include 14 *Avenger*-class MCMs, 17 MHCs, and 32 advanced MH–53 helicopters.

Leader and *Avenger* in the Persian Gulf, 1991. Desert Storm MCM operations required heroic efforts of MSO and MCM crews to clear gunfire support areas and to reopen shipping channels to Kuwait.

Although stringent budget cuts anticipated throughout the Navy in the 1990s threaten the completion of this modest building program, the entire cost of all active and reserve U.S. Navy funding for mine warfare, including the building of these new ships, accounts for less than one-half of one percent of the U.S. Navy budget.[76]

Recent concern over mines in the Persian Gulf has benefitted the MCM force by reemphasizing the urgent need for new ships. Already some of the new MCMs under construction, slated for turnover to the reserves after one year of operation, will now be retained in the active force. With new ships will come increased officer and enlisted billets. Also recognizing the need to better integrate MCM back into naval warfighting, in 1988 CNO Admiral Trost dual-hatted the Charleston-based COMINEWARCOM with an OPNAV collateral duty position as Director Mine Warfare Division (OP–72) under the DCNO for Naval Warfare, giving COMINEWARCOM a Washington base from which to champion mine warfare readiness. Yet, in 1991, in the midst of the biggest MCM operation since 1952, the COMINEWARCOM billet was temporarily gapped, leaving mine warfare once again a collateral duty of Commander Naval Base, Charleston.

Conclusion

Over the past two centuries the United States has failed to sustain an adequate capability in naval mine countermeasures, particularly in comparison to its other capabilities. Observing the continuing naval tradition of peacetime neglect of mine matters, Rear Admiral Brian McCauley predicted in 1973 that "rarely will anyone in today's Navy argue against the effectiveness of mine warfare nor our vulnerability as a nation to its use by other powers. Yet the practical demise of the Mine Force in the U.S. Navy is already planned—a victim of other more sophisticated higher priority programs."[1] Eight years later the General Accounting Office reported to Congress that "the Navy lacks the ability to lay mines in seas or harbors and is also short of the personnel and equipment needed to counter enemy mining. The Navy would find it hard to conduct even the most limited type of mining or mine countermeasures operation."[2] In 1985 Vice Admiral Joseph E. Metcalf, who had not been a previous supporter of the mine warfare forces, testifying before Congress on the subject of Soviet mine warfare, stated that "no element of our Navy is as deficient in capability against the threat as is the mine countermeasures force."[3] And in 1989 CNO Admiral Carlisle A. H. Trost observed that "until recently, the United States has not given enough sustained attention to maintaining a superior capability in mine warfare, particularly mine countermeasures. . . . I intend to keep attention focused on our vulnerability, and continue to press for resources to put us in a position where we can adequately protect our interests and deter potential adversaries."[4]

The reasons for this situation are complex. Although many defense observers have noted that "mine countermeasures have never had a strong constituency in the U.S. Navy," their explanations for this lack of support have varied. Some analysts have explained that most naval officers are vehicle-oriented rather than weapon-oriented and tend to cling to the communities behind the weapons systems, be they air, surface, or subsurface. Analysts further explain that since all major warfare disciplines are responsible for mine warfare assets, no one warfare specialty takes the responsibility to fund and support them seriously. Others say that the traditional secrecy and lack of widespread knowledge regarding mine warfare matters has exacerbated their low visibility and lower priority within the Navy. Still others point out that mine warfare continues to be unglamorous and that the community is simply considered by many naval officers to be a defensive backwater of modern naval operations.[5]

Competing for scarce resources with other Navy programs that have higher priorities, mines and MCM have most often been the losers. Navies, however, are composed of people, and people often base decisions on perceptions of the magnitude of the risk. Historically, the Navy has quite correctly associated the development of only minimal MCM capability as less risky than limiting other warfare areas.

Every Navy seeks a balanced fleet, but the makeup of that balance depends on the perception of the threat. The U.S. Navy has had no need to clear enemy mines from its own territorial waters since 1942. All later MCM operations have taken place overseas and have been generally limited and localized. Relatively few wartime operations of the U.S. Navy have been seriously affected by the presence of mines, and few Americans have expressed qualms about our continuing reliance on our NATO allies' minesweeping capabilities.

Nations like Britain and the Soviet Union, however, which may have a far better sense of their own history, place more value on training and retaining their MCM personnel. These nations use their existing technology and carefully accumulated knowledge to advance their naval forces. More important in the days of *glasnost* and *perestroika*, recent escalation of low-intensity mine warfare, particularly in the Middle East, has clearly demonstrated that Third World nations can easily obtain and use mines to stop naval and maritime forces.[6]

Most often the lessons learned by the operational MCM experience of the U.S. Navy have been forgotten, misinterpreted, or simply misapplied. Early observers of Robert Fulton's unsuccessful mining attempts believed that mines were unnecessary devices and that countermeasures could be easily fabricated when needed. Nothing in the Navy's nineteenth-century MCM experience convinced the naval establishment otherwise. As long as mines required contact with hulls, and torpedoes and mines remained inextricably linked in research and development applications, torpedoes, as high-tech weapons, received most of the attention and funding of the ordnance establishment. Until the Russians and the Japanese operationally proved the effectiveness of low-tech mining of the open sea in the early twentieth century, the U.S. Navy paid little attention to mine warfare and gave even less encouragement to the development of new MCM methods.

Over the course of the twentieth century offensive use of mines increased in importance to the Navy. Yet low casualty rates to mines and the apparently easy success of clearance efforts during both world wars wrongly convinced many Americans that countering mines remained a relatively easy task. The availability of allied technology and the use of jury-rigged equipment allowed the Navy to respond to wartime requirements without seriously advancing its MCM capability during peacetime.

The lessons learned from successful World War II MCM operations regarding our real vulnerabilities to advanced influence mines quickly became diluted by continuing budget crises, competing priorities within the

Stuart Carlson's 1987 cartoon from the *Milwaukee Sentinel*.

Navy and the Department of Defense, and the changing focus of strategic planning in the postwar period. Until mixed mine fields at Wonsan embarrassed the Navy and directed the attention of the nation toward the U.S. Navy's limited funding for MCM capabilities, MCM remained a stepchild of mine warfare, itself a stepchild of the total force.

By applying the operationally tested principles of MCM after Wonsan, the Navy developed a capable MCM fleet and community yet found little use for them in peacetime. Diverted by competing needs and the development of new strategic objectives, and drained of financial resources, the Navy scrapped most of the surface MCM fleet, reduced mine force assets, established no long-term shipbuilding program replacements, and rushed AMCM from infancy to adulthood. As a result, when mines in the Persian Gulf in 1987 disrupted free transit in international waters, the Navy found that effective clearance required considerable funding and full commitment of all available MCM forces. The Navy had not taken to heart the main lessons learned by the MCM force throughout its history, namely, that minesweeping is tedious, minehunting is more tedious, and countering mines cannot be made easy, cheap, or convenient. All these activities, however, are essential tools of modern warfighting.

Mines are now far more complex than the elementary mechanisms once easily foiled by simple mechanical minesweeping. All contact mines and many influence types are readily available for sale to nearly any country that wishes to use them. Mine casings can now be manufactured of nearly any material and can be made to resemble rocks or bottom debris. Mines laid in silty waters can be buried quickly and are almost impossible to find. Developments in microprocessor-controlled mines allow tracking and targeting of specific ships. Furthermore, rising mines have made even very deep waters mineable and place our most expensive assets at high risk.[7]

Since the increase in mining incidents in the 1980s, nearly every nation with valuable ships to protect, including the United States, has invested in MCM. The Japanese Navy, forced to apply daily its lessons learned while sweeping American combination mines since World War II, have many trained officers at command level who know the intricacies of advanced MCM and who have used their experience to construct capable ships. German and Italian experiments with MCM-related shipbuilding methods and their tactical employment have greatly advanced minehunting techniques. Mine warfare in the Royal Navy has been a viable career profession since the World War I and, in the Soviet Navy, since World War II. The proliferation of nuclear weapons among nations has also given rise to the concept of deterrence: as long as strategic use of nuclear weapons is considered suicide by all parties, the use of more conventional weapons, including mines, will continue, if not increase. Navies can never afford to ignore the threat of naval mines.

The only really effective counter to mines is to prevent them from being laid. Joint service and Navy-wide initiatives, a forward deployment strategy, and increased intelligence-gathering can limit widespread mining attacks but cannot always prevent low-intensity mining. Once mines are in the water, the only practical solution is a flexible, balanced MCM force and increased attention to mine avoidance throughout the Navy. As one aviator discovered,

> We clearly need a rebirth of innovative thinking about both mining and mine countermeasures. The global aviation community has learned the hard way that ordnance and their countermeasures are best deployed in concert; about mines, torpedoes, and *their* countermeasures, however, we have not only insufficient cooperation, but virtual hostility.[8]

The MCM community has always existed in a potential leadership vacuum for future operations. Even during wartime, when MCM asserted itself as a necessary warfighting skill, the lack of professional naval officers experienced in mine warfare has led to regular recalls from retirement and dual operational and administrative tasking in times of crisis.[9] Even now most of the generation of mine warfare specialists trained in the flush years after Wonsan have retired. Civil service and contract engineers sometimes find MCM as little career enhancing for civilians as it is for naval personnel.

Historically, the best young officers in MCM command and leadership positions have left the community to seek promotion or watched their careers stall. With billets being created for the new ships and aircraft, the mine force has a unique opportunity to begin rebuilding the MCM community effectively enough to make it last.

To have an adequate peacetime MCM proficiency capable of wartime expansion, a navy, any a navy, has to have sufficient trained personnel. As Rear Admiral Horne explained, "The new mine warfare hardware will not realize its potential unless we identify, develop, motivate and retain personnel who are mine warfare qualified by experience and performance." If a dynamic, identifiable leadership cadre can be developed to direct rebuilding of the Navy's MCM program to the level of flexible response capability that it reached from 1971 to 1975, history does not have to be repeated.

Current MCM leadership billets do not provide such direction for the MCM force and have not done so since the disestablishment of MINEWARFOR in 1975. The billet of Commander Mine Warfare Command has not always been promptly filled, and those who do fill it often find that they must spend all of their energies politically and actively reminding the Navy of its need for the warfighting skills of a properly equipped and manned MCM force. Given the lack of mine and MCM consciousness of the Navy as a whole, this task sometimes requires substantial assertiveness. As one observer noted, "until we get somebody in that force who is willing to be obnoxious, . . . Until we get somebody who is willing to make people in Washington wake up to the seriousness of this problem, at risk of life, limb, and career, we're going to be in the same boat 15 years from now."[10] Even with such aggressive mine warfare leadership, COMINEWARCOM's strictly advisory role limits his ability to assist the Navy in reintegrating MCM back into naval warfare. Even Admiral Zumwalt suggested in the 1970s that the problems associated with direction of mine warfare readiness required creation of a vice admiral billet in the Pentagon for mine warfare—an idea with which George Dewey would probably have agreed.[11]

The central problem of MCM throughout history has been the difficulty of sustaining maximum capability over time. By its very nature MCM evolves as the result of new mine developments and changing threats. Yet, in the U.S. Navy MCM have often been quick-fix solutions. Due to real competing needs, priorities, and lack of mine warfare knowledge within the Navy, it has been impossible to sustain adequate priority and funding for MCM. Important lessons learned, even when published by the participants, have been quickly forgotten, and subsequent attempts to revitalize the service have often been predicated on the wrong lessons. To date, no Chief of Naval Operations, Congress, or President has been opposed to an effective mine warfare program, and some have actively championed one. Yet, without historical perspective, recurring attempts to find an answer to the problem of an adequate MCM capability will continue to fail.

Lack of overall mine consciousness has often led us to remember the wrong lessons from our mine warfare experience. The recent experiences of *Samuel B. Roberts*, *Tripoli*, and *Princeton* remind us that even our most valuable and expensive warships can be easily stopped by simple, cheap mines. When the Navy as a whole learns more about the reality and potential of mines and their countermeasures, MCM will no longer be called the Cinderella of the service and considered a subject about which much is written and less done. Only knowledge will end the legends and reveal the truth about men like Farragut who only "damned" the torpedoes by actively hunting them to determine the risk.

Abbreviations

AM	Fleet Minesweeper
AMC	Coastal Minesweeper
AMCM	Airborne Mine Countermeasures
AMCU	Coastal Minesweeper (Underwater Locator)
AMS	Auxiliary Motor Minesweeper
APD	High-speed Transport
APL	Barracks Craft (non-self-propelled)
ARS	Salvage Ship
BUORD	Bureau of Ordnance
CINCPACFLT	Commander in Chief, Pacific Fleet
CMEF	Commander Middle East Force
CNO	Chief of Naval Operations
COMINCH	Commander in Chief, U.S. Fleet
COOP	Craft of Opportunity
CTG	Commander Task Group
DCNO	Deputy Chief of Naval Operations
DLC	Library of Congress
DMS	Destroyer Minesweeper
DMZ	Demilitarized Zone
DNA	National Archives and Records Administration
DOD	Department of Defense
EOD	Explosive Ordnance Disposal
HASC	House Armed Services Committee
HMH	Helicopter, Marine, Heavy
HMM	Helicopter, Marine, Medium
JMS	Japanese Motor Minesweeper

JTFME	Joint Task Force, Middle East
LAMPS	Light Airborne Multi-Purpose System
LCIL	Landing Craft, Converted, Steel-hulled
LCM	Landing Craft, Mechanized
LCPL	Landing Craft, Personnel, Large
LCVP	Landing Craft, Vehicle and Personnel
LPD	Amphibious Transport
LPH	Amphibious Assault Ship
LSD	Landing Ship, Dock
LST	Landing Ship, Tank
MCM	Mine Countermeasures Ship
MCMGRUCOM	Mine Countermeasures Group Command
MCS	Mine Countermeasures Command Ship
MHC	Minehunter, Coastal
MINELANT	Mine Force, Atlantic Fleet
MINEPAC	Mine Force, Pacific Fleet
MINEWARCOM	Mine Warfare Command
MINEWARFOR	Mine Warfare Force
MINRON	Mine Squadron
MLMS	Minesweeping Launch
MNS	Mine Neutralization System
MOMAG	Mobile Mine Assembly Group
MOMCOM	Mobile Mine Countermeasures Command
MOP	Magnetic Orange Pipe
MSB	Minesweeping Boat
MSC	Minesweeper, Coastal
MSD	Minesweeper, Drone
MSH	Minesweeper, Hunter
MSI	Minesweeper, Inshore

MSL	Minesweeping Launch
MSM	River Minesweeper
MSO	Minesweeper, Ocean
MSR	Patrol Minesweeper
MSS	Minesweeper, Special
MSSB	Minesweeping Shrimp Boat
NAVAIRLANT	Naval Air Forces, Atlantic Fleet
NAVFE	Naval Forces, Far East
NAVSEA	Naval Sea Systems Command
NAVSURFLANT	Naval Surface Force, Atlantic
NAVSURFPAC	Naval Surface Force, Pacific
NCSC	Naval Coastal Systems Center, Panama City, FL
NHC	Naval Historical Center, Washington, DC
NOL	Naval Ordnance Laboratory
OA	Operational Archives
ONRL	Office of Naval Records and Library
OPNAV	Office of the Chief of Naval Operations
PBM	Patrol Bomber
PBR	Patrol Boat, River
PCE	Patrol Craft, Escort
PCS	Patrol Craft, Submarine
PINS	Precise Integrated Navigational System
RG	Record Group
ROK	Republic of Korea
ROV	Remotely Operated Vehicle
SECNAV	Secretary of the Navy
SERVRON	Service Squadron
SH	Ships Histories Branch
UDT	Underwater Demolition Team

USCENTCOM	U.S. Central Command
USCINCCENT	U.S. Commander in Chief, Central Command
XMAP	Experimental Magnetic, Acoustic, and Pressure Sweep
YMS	Motor Minesweeper
YR	Floating Workshop (non-self-propelled)
YRBM	Repair, Berthing, and Messing Barge

Notes

Introduction

1. Log of USS *Hartford*, 31 Jul 1864, Record Group (RG) 24, National Archives (DNA), Washington, DC; Wilfrid Bovey, "Damn the Torpedoes...?", U.S. Naval Institute *Proceedings* (hereafter *Proceedings*) 65 (Oct 1939): 1446. The term "torpedo," named for the electric ray fish that shocks its prey into submission, was first used to describe land as well as sea mines, and continued to be used interchangeably to describe most underwater ordnance well into the twentieth century. Specialists in undersea warfare began distinguishing between stationary mines and automotive torpedoes as the British Whitehead torpedo gained popularity in the early 1870s, but the distinction gained popular application slowly. By the turn of the century both the warheads inside torpedoes and the explosives inside mine casings were still popularly known as "torpedoes." See R. B. Bradford, *History of Torpedo Warfare* (Newport, RI, 1882), pp. 3, 68, 90.

2. Virgil Carrington Jones, *The Civil War at Sea, July 1863–November 1865: The Final Effort* (New York, 1962), pp. 238–39, 417–21; Milton F. Perry, *Infernal Machines: the Story of Confederate Submarine and Mine Warfare* (Baton Rouge, 1965), pp. 157–59; Bovey, "Damn the Torpedoes," p. 1447; John Coddington Kinney, "Farragut at Mobile Bay," *Battles and Leaders of the Civil War* (New York, 1884), 4:382, 388; Franklin Matthews, *Our Navy in Time of War* (New York, 1899), p. 101; Farragut to Welles, 25 May 1864, quoted in U.S. Navy Department, *Official Records of the Union and Confederate Navies in the War of the Rebellion* (hereafter *ORN*) (Washington, 1894–1913), series 1, vol. 21 , p. 298; Log of USS *Cowslip*, 3 Jul 1864 entries, RG 24, DNA. See also Farragut to T. Bailey, 26 May 1864, in *ORN*, 1, 21:298–99.

3. *Hartford* log, 30 Jun, 25 and 27 Jul 1864; Log of USS *Sebago*, 25 Jul 1864, RG 24, DNA; Alexander McKinley Diary, 25 Jul 1864, Hagley Museum and Library, Wilmington, DE; Perry, *Infernal Machines*, p. 159; Jones, *Civil War at Sea*, p. 242; *Cowslip* log, 3 and 4 Jul 1864, Log of USS *Glasgow*, 21 Jul 1864, RG 24, DNA; Viktor Ernst Karl Rudolf von Scheliha, *A Treatise on Coast-Defence . . .* (London, 1868), pp. 225–29.

4. John C. Watson, "Farragut and Mobile Bay—Personal Reminiscences," *Proceedings* 53 (May 1927): 555–56; J. C. Watson, "Some of my personal recollections of Admiral Farragut. For the information of Captain A. T. Mahan, U.S.N., in writing a biography of the Admiral," John Crittenden Watson Papers (hereafter Watson Papers), Library of Congress, Washington, DC (hereafter DLC).

5. Jones, *Civil War at Sea*, pp. 243–45; Remarks of Rear Admiral C. M. Chester, in John C. Watson, "Farragut and Mobile Bay—Personal Reminiscences," *War Paper No. 98* (Paper delivered at the meeting of the Military Order of the Loyal Legion of the United States, Commandery of the District of Columbia, 16 Dec 1916), pp. 62–63; Log of US Monitor *Manhattan*, 2 Aug 1864; *Hartford* log, 2 and 3 Aug 1864, RG 24, DNA; Charles Lee Lewis, *David Glasgow Farragut: Our First Admiral* (Annapolis, 1943), p. 244; Percival Drayton to Alexander Hamilton, Jr., 19 Aug 1864, in *Naval Letters From Captain Percival Drayton, 1861–1865* (New York: New York Public Library, 1906), p. 67. Watson's 2 August attempt was unsuccessful. *Manhattan* log, 3 Aug 1864.

6. General Orders No. 11, 29 Jul 1864, *ORN*, 1, 21:398.

7. Journal of J. C. Gregg, USS *Brooklyn*, 25 Jul and 4 Aug 1864 entries, RG 45, DNA; Farragut to T. A. Jenkins, 3 Aug 1864, *ORN*, 1, 21:403; Watson, "Farragut and Mobile Bay," *Proceedings*, p. 553; William F. Hutchinson, "The Bay Fight: A Sketch of the Battle of Mobile Bay, August 5th, 1864," *Personal Narratives of the Battles of the Rebellion. Being Papers Read Before the Rhode Island Soldiers and Sailors Historical Society. No. 8* (Providence, RI: Sydney S. Rider, 1879), p.

12; Charles R. Haberlein, Jr., "Damn the Torpedoes!" in William C. Davis, ed., *The Image of War, 1861–1865*, vol. 6, *The End of an Era* (Garden City, NY, 1984), pp. 86–96. Although it is generally believed that a keg torpedo was responsible, a Confederate torpedo officer later claimed that the mine which sunk *Tecumseh* was electrically fired from Fort Gaines. John Sanford Barnes, *Submarine Warfare, Offensive and Defensive* (New York, 1869), p. 108.

8. Watson, "Farragut and Mobile Bay—Personal Reminiscences," in *War Paper 98*, p. 13. In several accounts bystanders related strikingly similar words. Watson, near Farragut, recalled his full quote as "Damn the Torpedoes! Full speed ahead, Drayton! Hard a starboard! Ring four bells! Eight bells! Sixteen bells!" Ibid., and Watson, "Farragut and Mobile Bay," *Proceedings*, p. 555. These comments underscore Farragut's anger and impatience as he seized the initiative and charged into the bay. See also Lewis, *David Glasgow Farragut*, p. 469. In a much later account Chief Engineer Thomas Williamson related that Farragut merely ordered "Go ahead," and in reply to Williamson's question, "Shall I ring four bells, sir?" to alert the engine room to full speed ahead, Farragut yelled "Four bells —eight bells—*sixteen* bells—damn it, I don't care how many bells you ring!" *Proceedings* 65 (Nov 1939): 1676. Despite Farragut's known aversion to swearing, he probably did say, "Damn the torpedoes."

9. Alexander McKinley to Samuel F. DuPont, 18 Sep 1864, in John D. Hayes, ed., *Samuel Francis DuPont: A Selection From His Civil War Letters*, vol. 3, *The Repulse: 1863–1865* (Ithaca: Cornell University Press, 1969), pp. 381–87; Perry, *Infernal Machines*, p. 161. The quote is from Watson, "Farragut and Mobile Bay," *Proceedings*, p. 555.

10. General Orders No. 12, 6 Aug. 1864, *ORN*, 1, 21:438. Artists and newspapers made much of the vision of Farragut lashed to *Hartford*'s rigging. Perhaps in reaction to this portrayal, which he saw in New York newspapers shortly after the battle, Farragut emphasized in his official report several days later the heroism of his fleet and the dangers it faced, further clouding the extent of Watson's minehunting efforts. "Though he had not been able to discover the sunken torpedoes," Farragut mistakenly claimed, "yet we had been assured by refugees, deserters, of their existence, but believing that, from their having been some time in the water, they were probably innocuous, I determined to take the chance of their explosion." Farragut to Welles, 12 Aug 1864, *ORN*, 1, 21:417. A month later, amused but still smarting from the humorous insinuations, Farragut denied to his wife that he had been imprisoned in the rigging. "The illustrated papers are very amusing. Leslie has me lashed up to the mast like a culprit & says that is the way officers will hereafter go into Battle. . . ." Farragut to wife, 15 Sep 1864, David G. Farragut Papers, Naval Historical Foundation (NHF), Washington, DC.

11. Remarks of Rear Admiral Chester in *War Paper No. 98*, p. 19.

12. Barrett to LT J. T. E. Andrews, 20 Aug 1864, *ORN*, 1, 21:569.

13. Farragut to Welles, 29 Aug 1864, *ORN*, 1, 21:616; Bradford, *History of Torpedo Warfare*, p. 58.

14. Office of the Chief of Naval Operations (OPNAV) Instruction (OPNAVINST) 3370.3B, ser 09/300616, 9 Jul 1982, Mine Warfare Policy.

15. Gregory Kemenyi Hartmann, *Weapons That Wait: Mine Warfare in the U.S. Navy* (Annapolis, 1979), pp. 94, 266–67.

16. Richards T. Miller, "Minesweepers," *Naval Review 1967* (Annapolis, 1966): 210; Testimony of VADM James R. Hogg, DCNO for Naval Warfare, *Mine Warfare: Hearing Before the Seapower and Strategic and Critical Materials Subcommittee of the Committee on Armed Services*, 100th Cong., 1st sess., 1987, H.A.S.C. 100–50 (hereafter HASC Hearing), pp. 19–24.

17. D. W. Feldman, "Historical Review of Mine Countermeasures Techniques and Devices," Technical Note No. 449, Mine Countermeasures Department, Naval Coastal Systems Center, Panama City, FL, p. 1.

1. A Matter of Efficacy

1. Bradford, *History of Torpedo Warfare*, p. 7.

2. J. S. Cowie, *Mines, Minelayers and Minelaying* (London, 1949), pp. 7–10.

3. Gregory K. Hartmann, "Mine Warfare History and Technology," Naval Surface Weapons Center, Technical Report No. 75–88, White Oak Laboratory, MD, 1 Jul 1975, p. 9; Robert Fulton, *Torpedo Warfare and Submarine Explosions* (New York, 1810), p. 7.

4. Bradford, *History of Torpedo Warfare*, pp. 18–19.

5. C. W. Sleeman, *Torpedoes and Torpedo Warfare: Containing a Complete and Concise Account of the Rise and Progress of Submarine Warfare; Also a Detailed Description of all Matters Appertaining Thereto, Including the Latest Improvements* (Portsmouth, UK, 1880), p. 6; Barnes, *Submarine Warfare*, pp. 41–42; Bradford, *History of Torpedo Warfare*, pp. 18, 20.

6. Cowie, *Mines, Minelayers and Minelaying*, pp. 11–15; Hartmann, *Weapons that Wait*, p. 29; J. N. Ferguson, "The Submarine Mine," *Proceedings* 40 (Nov-Dec 1914): 1698.

7. Ferguson, "The Submarine Mine," p. 1699; Bradford, *History of Torpedo Warfare*, pp. 30–31. See also Philip K. Lundeberg, *Samuel Colt's Submarine Battery: The Secret and the Enigma* (Washington, 1974).

8. Bradford, *History of Torpedo Warfare*, pp. 35–36; Harry William Edwards, "A Naval Lesson of the Korean Conflict," *Proceedings* 80 (Dec 1954): 1338; Philip K. Lundeberg, "Undersea Warfare and Allied Strategy in World War I. Part I: to 1916," *Smithsonian Journal of History* 1 (Autumn 1966): 4; Perry, *Infernal Machines*, p. 4; Robert C. Duncan, *America's Use of Sea Mines* (Silver Spring, MD, 1962), pp. 41–42.

9. Bradford, *History of Torpedo Warfare*, pp. 37–39, 40.

10. Arnold S. Lott, *Most Dangerous Sea: A History of Mine Warfare and an Account of U.S. Navy Mine Warfare Operations in World War II and Korea* (Annapolis, 1959), 25 Mar 1864, p. 10.

11. Nathan Reingold, "Science in the Civil War: The Permanent Commission of the Navy Department," *ISIS* 49 (Sep 1958): 317. The main concerns of Welles's Naval Examining Board, like most of the Navy establishment, concerned magnetism, steam motive power, and armored ships rather than ordnance. Ibid., p. 315.

12. Barnes, *Submarine Warfare*, p. 181.

13. Alex Roland, *Underwater Warfare in the Age of Sail* (Bloomington, 1978), p. 162.

14. Perry, *Infernal Machines*, pp. 5–27, 31.

15. Scheliha, *A Treatise on Coast-Defence*, pp. 223–25, 231–35; Perry, *Infernal Machines*, pp. 37–38, 43–45.

16. Perry, *Infernal Machines*, pp. 35–36, 56, 117; Scheliha, *A Treatise on Coast-Defence*, p. 221; Barnes, *Submarine Warfare*, pp. 180–81.

17. Bradford, *History of Torpedo Warfare*, pp. 47, 50.

18. Perry, *Infernal Machines*, pp. 33, 34; David Dixon Porter to Gideon Welles, 17 Dec 1862, in Barnes, *Submarine Warfare*, p. 82; Thomas O. Selfridge to Henry Walke, 13 Dec 1862, ibid., p. 80; Edwin C. Bearss, *Hardluck Ironclad: The Sinking and Salvage of the Cairo* (Baton Rouge, 1966), pp. 97–99. In the wake of *Cairo*'s loss Union boats searched the waters and found and destroyed twelve mines nearby. Selfridge had three ships, *Cumberland*, *Cairo*, and *Conestoga*, sunk from under him during this war.

19. Barnes, *Submarine Warfare*, p. 181; Wm. C. Hanford to D. D. Porter, 5 Sep 1864, *ORN*, 1, 26:552.

20. General Order No. 75, U.S. Mississippi Squadron, 24 Jul 1863, *ORN*, 1, 25:320–21; Wm. C. Hanford to Porter, 5 Sep 1864, ibid., 1, 26:552–53; Perry, *Infernal Machines*, pp. 45, 48.

21. Perry, *Infernal Machines*, pp. 51–52.

22. Barnes, *Submarine Warfare*, p. 87; Perry, *Infernal Machines*, pp. 57, 109, 113, 148–49, 168, 208.

23. Journal of J. C. Gregg, 25 Aug 1864 entry; Perry, *Infernal Machines*, pp. 162, 182–87; Miller, "Minesweepers," p. 210.

24. Bradford, *History of Torpedo Warfare*, p. 47; Cowie, *Mines, Minelayers and Minelaying*, p. 20; Perry, *Infernal Machines*, pp. 175, 179, 191–92. After the war Maury and his mine systems went farther south where they both saw service in Mexico.

25. J. C. Beaumont to S. P. Lee, 6 May 1864, *ORN*, 1, 10:9; John S. Barnes to Lee, 10 May 1864, ibid., pp. 10–11; Thomas F. Wade to Welles, 13 May 1864, ibid., pp. 14–15; Perry, *Infernal Machines*, pp. 111–13. These effective controlled mines had been in the water nearly two years.

26. Perry, *Infernal Machines*, pp. 148–49.

27. Ibid., pp. 172–73, 175, 180-81.

28. Bradford, *History of Torpedo Warfare*, pp. 46–47. For a list of mined Union vessels see Perry, *Infernal Machines*, pp. 199–201.

29. Barnes, *Submarine Warfare*, p. 181; Hanford to Porter, 5 Sep 1864, *ORN* 1, 26:552. Young Acting Volunteer Lieutenant George W. Brown devised an anchor buoy to protect his ship, *Ozark*, on the Mississippi River. When the buoy successfully snagged a Confederate mine, Porter forwarded Brown's design to the Navy Department for consideration, but Welles showed no interest in the invention or in passing the news to the fleet. Brown to Porter, 19 Sep 1864; Porter to Welles, 28 Sep 1864, *ORN*, 1, 26:569

30. Bradford, *History of Torpedo Warfare*, pp. 42, 77–81, 83–87; Perry, *Infernal Machines*, p. 197.

31. Bernard Brodie, *Sea Power in the Machine Age* (Princeton, 1941), p. 276; Feldman, "Historical Review of Mine Countermeasures," p. 13; 14; Cowie, *Mines, Minelayers and Minelaying*, pp. 27–28; Ropp, *The Development of a Modern Navy*, p. 178.

32. A. Ludlow Case to Geo. M. Robeson, 23 Oct 1869, U.S. Navy Department, *Report of the Secretary of the Navy, Showing the Operations of the Department for the Year 1869* (Washington, 1869) (all departmental reports hereafter SECNAV *Annual Report*), pp. 68, 71; Case to Robeson, 19 Oct 1870, ibid., 1870, p. 55; Porter to Robeson, 10 Nov 1870, ibid., p. 163; Case to Robeson, 17 Oct 1872, ibid., 1872, p. 52; Bradford, *History of Torpedo Warfare*, p. 89.

33. John Rodgers et al to Case, 31 Jul 1872, SECNAV *Annual Report*, 1872, p. 55.

34. Porter to Robeson, 22 Oct 1873, SECNAV *Annual Report*, 1873, pp. 276–79.

35. Porter to Robeson, 7 Nov 1874, SECNAV *Annual Report*, 1874, pp. 207–8; Sleeman, *Torpedoes and Torpedo Warfare*, p. 239; William N. Jeffers to Robeson, 2 Oct 1876, SECNAV *Annual Report*, 1876, p. 113.

36. Jeffers to R. W. Thompson, 10 Oct 1878, SECNAV *Annual Report*, 1878, p. 66; William H. Hunt to the President, 28 Nov 1881, ibid., 1881, p. 15; Bradford, *History of Torpedo Warfare*, pp. 54–55.

37. Montgomery Sicard to Chandler, 1 Nov 1882, SECNAV *Annual Report*, 1882, 3:8; Selfridge to Sicard, 29 Sep 1882, ibid., pp. 18–24; Sicard to Chandler, 1 Nov 1883, ibid., 1883, p. 414; Selfridge report, ibid., p. 425; Sicard to Chandler, 10 Nov 1884, ibid., 1884, pp. 418, 420.

38. John B. Hattendorf, B. Mitchell Simpson III, and John R. Wadleigh, *Sailors and Scholars: The Centennial History of the U.S. Naval War College* (Newport, 1984), pp. 18, 27; Ronald Spector, *Professors of War: The Naval War College and the Development of the Naval Profession* (Newport, 1977), pp. 55–58; W. T. Sampson to Sicard, n.d., SECNAV *Annual Report*, 1885, p. 248; Ruddock F. Mackay, *Fisher of Kilverstone* (Oxford, 1973), p. 377.

39. Brodie, *Sea Power*, p. 279; Bradford, *History of Torpedo Warfare*, pp. 87–89; Spector, *Professors of War*, p. 5; Ropp, *The Development of a Modern Navy*, p. 305; Porter to Chandler, 29 Nov 1882,

SECNAV *Annual Report*, 1882, 1:243; "Torpedo Work in Great Britain, France, and Norway," ibid., 3:44–70.

40. Porter to William C. Whitney, 15 Nov 1886, SECNAV *Annual Report*, 1886, p. 61; C. F. Goodrich to Chief of the Bureau of Ordnance (BUORD), 13 Oct 1887, ibid., 1887, pp. 266–67; Porter to SECNAV, 18 Jul 1888, ibid., 1888, p. 12.

41. It was not until 1976 that scientists and engineers established conclusively that *Maine* was destroyed not by a mine but by internal implosion. See H. G. Rickover, *How the Battleship Maine Was Destroyed* (Washington, 1976).

42. T. C. McLean to Chief BUORD, 8 Sep 1898, SECNAV *Annual Report*, 1898, 1:499; Buford Rowland and William B. Boyd, *U.S. Navy Bureau of Ordnance in World War II* (Washington, n.d.), p. 157.

43. George Dewey, *Autobiography of George Dewey, Admiral of the Navy* (New York, 1913), pp. 91, 117–211. Dewey served as first lieutenant on *Mississippi* under Captain Melancthon Smith.

44. Ibid., pp. 198–99; Ronald Spector, *Admiral of the New Empire: The Life and Career of George Dewey* (Baton Rouge, 1974), pp. 49–53.

45. Joseph L. Stickney, *Admiral Dewey at Manila* (Philadelphia, PA, 1899), p. 35; Richard S. West, *Admirals of American Empire* (New York, 1948), pp. 201–2.

46. Dewey, *Autobiography*, p. 202; H. W. Wilson, *The Downfall of Spain: Naval History of the Spanish-American War* (Boston, 1900), pp. 125–26, 158–59; Spector, *Admiral of the New Empire*, p. 54.

47. George Edward Graham, *Schley and Santiago* (Chicago, 1902), pp. 112–23; Wilson, *Downfall of Spain*, pp. 260, 398.

48. W. A. M. Goode, *With Sampson Through the War* (New York, 1899), pp. 145–46; David F. Trask, *The War With Spain in 1898* (New York, 1981), pp. 135–36. Although Sampson did not know it, the mines consisted of only two lines, one of 6 and one of 7 mines, each separately controlled.

49. Cowie, *Mines, Minelayers and Minelaying*, pp. 30–31; Admiral Sampson to General Shafter, Jun 1898, quoted in CAPT J. H. A. Day, USMC, "Discussion on the Use of Mines," 2 Oct 1911, General Board Study No. 431, pp. 21–22.

50. Wilson, *Downfall of Spain*, pp. 174–75, 382.

51. Hartmann, "Mine Warfare History and Technology," p. 6.

52. George Dewey to SECNAV, 18 Feb 1901 and endorsements, ser 431, Records of the General Board 1913–1931, RG 80, DNA.

53. U.S. Naval Torpedo Station, *Mines and Countermines, U.S.N.* ([Newport, RI], 1902), pp. 49–68.

54. Cowie, *Mines, Minelayers and Minelaying*, pp. 33–35; Ellis A. Johnson and David A. Katcher, *Mines Against Japan* (Washington, 1973), p. 2; Day, "Discussion on the Use of Mines", p. 2; C. C. Wright, "Origins of the 'Bird' Class Minesweeper Design," appendix to Harry C. Armstrong, "The Removal of the North Sea Mine Barrage," *Warship International* 25 (Jun 1988): 164; CAPT E. H. Ellis, USMC, "Discussion on the Use of Mines," 2 Oct 1911, General Board Study No. 431, p. 2. For the effects of mine warfare in this war, see M. C. Ferrand, "Torpedo and Mine Effects in the Russo-Japanese War," translated by Philip R. Alger, *Proceedings* 33 (Dec 1907): 1482–85.

55. N. E. Mason to SECNAV, 1 Oct 1905, SECNAV *Annual Report*, 1905, pp. 13, 31; Duncan, *America's Use of Sea Mines*, p. 36.

56. Mackay, *Fisher of Kilverstone*, pp. 377–78; Hartmann, "Mine Warfare History and Technology," p. 14; Cowie, *Mines, Minelayers and Minelaying*, pp. 35–36, 43; CDR G. R. Marvell, "Mines and Mining," 3 Jan 1916, General Board Study No. 431, RG 80, DNA; Ellis, "Discussion on the Use of Mines," p. 7.

57. Gordon Bruce, "Mine Sweeping," in Jane Anderson and Gordon Bruce, *Flying, Submarining and Mine Sweeping* (London, 1916), pp. 34–35; Cowie, *Mines, Minelayers and Minelaying*, pp. 168–80; Hartmann, "Mine Warfare History and Technology," p. 12; Henry Taprell Dorling, *Swept Channels: Being an Account of the Work of the Minesweepers in the Great War* (London, 1935), p. 7. Neither Russia nor Korea signed this Convention.

58. Cowie, *Mines, Minelayers and Minelaying*, p. 41; Marvell, "Mines and Mining."

59. Dewey, Memo to SECNAV, 30 Jan 1912, General Board No. 431, RG 80, DNA; Dewey to SECNAV, 30 Oct 1912, ibid.; G. V. L. Meyer to General Board, 12 Dec 1912 and 2 Jan 1913, ibid.; General Board (Dewey) to SECNAV, 2 Jan 1913, endorsed "approved" by Meyer, ibid.; Wright, "Origins of the 'Bird' Class," pp. 164, 168.

60. Wright, "Origins of the 'Bird' Class," pp. 164–65.

61. Cowie, *Mines, Minelayers and Minelaying*, pp. 46–48, 52–53; Archibald Montgomery Low, *Mine and Countermine* (New York, 1940), p. 190; "Mine Sweeping," *Proceedings* 42 (Jul-Aug 1916): 1311.

62. Wright, "Origins of the 'Bird' Class," pp. 164–165, 168; *Navy Register*, 1 Jan 1916, p. 275.

63. "Mine Sweeping Experiences," *Proceedings* 43 (Jun 1917): 1343–45; Jeffrey D. Wallin, *By Ships Alone: Churchill and the Dardanelles* (Durham, NC, 1981), pp. 152, 197; Henry W. Nevinson, *The Dardanelles Campaign* (New York, 1919), pp. 60–62; Andrew Patterson, Jr., "Mining: A Naval Strategy," *Naval War College Review* 23 (May 1971): 54.

64. Robert William Love, Jr., ed., *The Chiefs of Naval Operations* (Annapolis, 1980), p. xvii; Lott, *Most Dangerous Sea*, p. 14; Destroyer Force, Atlantic Fleet, USS *Seattle*, Flagship, 18 May 1917, "Instructions for Detection and Destruction of Floating Mines, . . . To be attached to Destroyer Force Doctrine," AC, Subject File 1911–27, RG 45, DNA.

65. M. A. Ransom, "The Little Gray Ships," *Proceedings* 62 (Sep 1936): 1281.

66. C. N. Hinkamp, "Pipe Sweepers," *Proceedings* 46 (Sep 1920): 1477-84.

67. Cowie, *Mines, Minelayers and Minelaying*, pp. 84–85, 105; Maurice Griffiths, *The Hidden Menace* (Greenwich, UK, 1981), pp. 61–63; Lott, *Most Dangerous Sea*, pp. 63–64.

68. Griffiths, *The Hidden Menace*, p. 61; Low, *Mine and Countermine*, p. 187.

69. Low, *Mine and Countermine*, p. 182; George L. Catlin, "Paravanes," *Proceedings* 45 (Jul 1919): 1135–57; condensed from George L. Catlin, "Paravanes: A History of the Activities of the Bureau of Construction and Repair in Connection With Paravanes in the War With Germany" (unpublished manuscript, 1919), pp. 1–11, Navy Department Library. Ensign, later Lieutenant, Catlin headed the Bureau of Construction and Repair's "Protective Devices" Section in the Maintenance Division and worked overseas with the British "Paravane Department." Ibid., pp. 6–8.

70. Catlin, "Paravanes," pp. 8, 24, 35; Low, *Mine and Countermine*, p. 188; Cowie, *Mines, Minelayers and Minelaying*, pp. 84–5, 105–6; Griffiths, *Hidden Menace*, pp. 63–65.

71. Griffiths, *Hidden Menace*, pp. 66–67; Cowie, *Mines, Minelayers and Minelaying*, p. 92.

72. Feldman, "Historical Review of Mine Countermeasures," p. 1314; Cowie, *Mines, Minelayers and Minelaying*, pp. 58, 9296, 103–104; Griffiths, *Hidden Menace*, pp. 67–69.

73. University of Pittsburgh, Historical Staff at the Office of Naval Research, "The History of United States Naval Research and Development in World War II" (unpublished manuscript, 1949) (hereafter ONR, "Research and Development"), p. 999, Navy Department Library; 23 Aug, 5 Sep, 21 Jun, 19 Jul, 8, 12, 14, and 16 Oct, 9, 21, 23 Nov entries, ZAG file, Enemy Mining Operations of the U.S. Coast, Subject File 1911–27, RG 45, DNA; Lott, *Most Dangerous Sea*, p. 16; Robert M. Grant, "The Use of Mines Against Submarines," *Proceedings*, 64 (Sep 1938): 1275.

74. Hartmann, *Weapons that Wait*, pp. 43, 48; Noel Davis, "The Removal of the North Sea Mine Barrage," *National Geographic* 37 (Feb 1920): 115; Lott, *Most Dangerous Sea*, p. 15; Cowie, *Mines, Minelayers and Minelaying*, pp. 65–66.

75. SECNAV *Annual Report*, 1919, p. 48; Armstrong, "Removal of the North Sea Mine Barrage," pp. 136–40; Hartmann, "Mine Warfare History and Technology," p. 18; Robert M. Grant, *U-boats Destroyed: The Effect of Anti-Submarine Warfare, 1914–1918* (London, 1964); J. A. Meacham, "The Mine as a Tool of Limited War," *Proceedings* 93 (Feb 1967): 60; Davis, "The Removal of the North Sea Mine Barrage," p. 105. Numbers of U-boats damaged and destroyed by the mine barrage are best documented by Grant, *U-Boats Destroyed*, pp. 99–108, 145–46. There is at present no source for the number of U-boats diverted by the presence of mines.

76. Memo, 26 Sep 1916, quoted in Wright, "Origins of the 'Bird' Class," p. 165.

77. Armstrong, "Removal of the North Sea Mine Barrage," p. 139.

78. Reginald Rowan Belknap, *The Yankee Mining Squadron or Laying the North Sea Mine Barrage* (Annapolis, 1920), p. 107; Wright, "Origins of the 'Bird' Class," pp. 166–67. After 1920 they were designated AM.

79. U.S. Navy Department, Office of Naval Records and Library (hereafter ONRL), *The Northern Barrage: Taking Up the Mines* (Washington, 1920), pp. 7–8; Davis, "The Removal of the North Sea Mine Barrage," pp. 103–7. SECNAV *Annual Report*, 1918, p. 395; ibid., 1919, p. 48; Armstrong, "Removal of the North Sea Mine Barrage," p. 139.

80. Armstrong, "Removal of the North Sea Mine Barrage," pp. 138–39; Report of RADM Joseph Strauss on Northern Sea Barrage, p. 9, enclosed with Strauss to CNO, 9 Dec 1919, AN: Northern Mine Barrage, Subject File 1911–27, RG 45, DNA.

81. Davis, "The Removal of the North Sea Mine Barrage," pp. 107–11; ONRL, *Northern Barrage: Taking Up the Mines*, pp. 10–11; Armstrong, "Removal of the North Sea Mine Barrage," p. 140. For a thorough report on the *Patapsco* and *Patuxent* experiments, see LT W. K. Harrill to Commander Mine Force [Strauss], 8 Dec 1918, box 6, AN: Northern Mine Barrage, Subject File 1911–27, RG 45, DNA.

82. Marvell, "Mines and Mining"; Feldman, "Historical Review of Mine Countermeasures," p. 8.

83. Armstrong, "Removal of the North Sea Mine Barrage," pp. 143–46.

84. ONRL, *Northern Barrage: Taking Up the Mines*, pp. 17–18, 26–30, 45; Davis, "The Removal of the North Sea Mine Barrage," pp. 111–18; Armstrong, "Removal of the North Sea Mine Barrage," pp. 144–50.

85. Armstrong, "Removal of the North Sea Mine Barrage," pp. 151–58; Duncan, *American Use of Sea Mines*, pp. 65–66; SECNAV *Annual Report*, 1919, pp. 48–49.

86. Armstrong, "Removal of the North Sea Mine Barrage," pp. 153, 158; SECNAV *Annual Report*, 1919, pp. 48–51; ONRL, *Northern Barrage: Taking Up the Mines*, pp. 43–45, 51, 58–61; Davis, "The Removal of the North Sea Mine Barrage," pp. 108, 113, 126–33; Hartmann, *Weapons that Wait*, p. 122; Strauss report, pp. 9–10; Strauss to VADM William S. Sims, Force Commander, 15 Dec 1918, box 5, AN: Northern Mine Barrage, Subject File 1911–27, RG 45, DNA; Harrill to Strauss, 8 Dec 1918, box 7, ibid.; Memo for Dudley Knox, 3 Mar 1938, ZAN: Northern Barrage, Subject File 1911–27, RG 45 DNA; SECNAV *Annual Report*, 1919, p. 51.

87. Wright, "Origins of the 'Bird' Class," p. 167; Jeffers to Robeson, 1 Nov 1873, SECNAV *Annual Report*, 1873, p. 103; Theo. F. Jewell to Chief BUORD, 1 Oct 1890, ibid., 1890, p. 274.

88. Ellis, "Discussion on the Use of Mines," General Board Study No. 431, 2 Oct 1911, p. 6.

89. Lott, *Most Dangerous Sea*, p. 17.

2. A New Menace

1. Richard W. Leopold, "Fleet Organization, 1919–1941" (unpublished manuscript), pp. 12–13, 18–19, Operational Archives, Naval Historical Center (hereafter OA).

2. Julius Augusta Furer, *Administration of the Navy in World War II* (Washington, 1959), p. 173; Leopold, "Fleet Organization," pp. 13–14, 19–20; Office of the Chief of Naval Operations, "Mine Warfare in the Naval Establishment" (hereafter CNO, "Mine Warfare") (unpublished manuscript), Washington, n.d., pp. 3–6, 10–11; Cowie, *Mines, Minelayers and Minelaying*, pp. 105–6. In 1935 all minesweeping and minelaying vessels were shifted to the administrative control of Commander Minecraft, Battle Force, reporting to the fleet commanders in chief. Mine Force, Pacific Fleet (MINEPAC), Command History, 1958, OA.

3. Hartmann, "Mine Warfare History and Technology," p. 21; *Conway's All the World's Fighting Ships, 1922–1946* (New York, 1980), p. 86; Ellsworth Dudley McEathron, "Minecraft in the Van: A History of United States Minesweeping in World War II," 3 vols., World War II Command File, OA.

4. S. D. Sturgis to Chief Umpire, 14 Jan 1924, Records Relating to United States Navy Fleet Problems 1 to 22, 1923–1941 (hereafter Fleet Problems), reel 3, RG 80, DNA; Commander Scouting Fleet to Commander in Chief, U.S. Fleet (CINCUS), 15 Feb 1924, ibid. (Fleet Problem 22, 1941 was cancelled because of the international situation.) The quote is from "Talk On Operations of Black Forces Delivered by Vice Admiral McCully before Conference on Problem No. 3, 21 January 1924," ibid.

5. CINCUS to Commander in Chief, Battle Fleet, 12 Oct 1926, Fleet Problem 7, reel 8; Blue Convoy Chronological History by Observer LCDR L. L. Babbitt, ibid; Black Course of Action, Fleet Problem 9, reel 12.

6. CINCUS to CNO, 15 Sep 1935, Comments on Phase 3, and General Comment, ibid., Fleet Problem 16, reel 18.

7. CNO to CINCUS, 9 Jul 1936, Fleet Problem 18, reel 22. Mines and gear were assumed to be present for the sake of the exercise.

8. CNO to CINCUS, 9 Jul 1936, Fleet Problem 18, reel 22; Commander Scouting Force to CINCUS, 2 Jun 1937, ibid., reel 23; Annex "B" to U.S. Fleet Op Order 2–40, 29 Mar 1940, Fleet Problem 21, reel 31. The mines in the Mine Barrage had been laid at depths to 950 feet. Mooring cables developed by 1937 allowed mining of depths to 3,000 feet, although they had not been operationally tested. Ibid. The 100-fathom assumption persisted, however, up to and including the last fleet problem exercise. Annex "B" to U.S. Fleet Op Order No. 2–40, 15 Jan 1940, Fleet Problem 21, reel 31.

9. Commander Scouting Force to CINCUS, 2 Jun 1937, Fleet Problem 18, reel 23.

10. Commander Minecraft, Battle Force, to COMINCH, 3 May 1940, ibid., reel 34.

11. Purple French Frigate Detachment (Rear Admiral Rowcliff) Remarks for the critique, 7 May 1940, Fleet Problem 21, reel 36.

12. Duncan, *America's Use of Sea Mines*, pp. 77, 85–86, 98–101; Hartmann, *Weapons That Wait*, pp. 57–58; Ellis A. Johnson and David A. Katcher, *Mines Against Japan*, (Washington, 1973), p. iii; Rowland and Boyd, *Bureau of Ordnance*, p. 159; Navy Department, U.S. Naval Administration in World War II, "History of the Naval Ordnance Laboratory, 1918–1945" (hereafter NOL Administrative History), part 1, pp. 3–7. NOL later moved from the Navy Yard to White Oak, Maryland, where it remains today.

13. "Countering the Magnetic Mine," *The Engineer* (15 Mar 1940), reprinted in *Proceedings* 66 (May 1940): 756; Hartmann, *Weapons That Wait*, p. 55; Lott, *Most Dangerous Sea*, pp. 67–68.

14. Jeffrey K. Bray, "Mine Warfare in the Russian and Soviet Navies" (M.A. Thesis, Old Dominion

University, 1989), pp. 95–97. During the war Germans also pulled up Russian mines, repositioning them to attack Soviet ships—a true example of offensive MCM. Ibid., pp. 91–92.

15. Cowie, *Mines, Minelayers and Minelaying*, pp. 100–102, 107, 120–21, 124; Hartmann, *Weapons That Wait*, p. 56; A. P. Lukin, "Secrets of Mine Warfare," *Proceedings* 66 (May 1940): 642–43; Peter Elliott, *Allied Minesweeping in World War II* (Annapolis, 1979), pp. 30–32, 40. For a description of the workings of the early magnetic mines, see C. E. Milbury, "Mystery of the Magnetic Mine," *Scientific American* (Mar 1940), reprinted in *Proceedings* 66 (May 1940): 754–56.

16. John Ennis, "Deep, Cheap & Deadly," *Sea Power* 17 (Dec 1974): 27. For a description of the British efforts to defeat the magnetic mine in 1940, see George Ashton, "Minesweeping Made Easy," *Proceedings* 87 (Jul 1961): 66–71.

17. Elliott, *Allied Minesweeping*, pp. 23–24, 35–36, 40, 49–50; Cowie, *Mines, Minelayers and Minelaying*, pp. 136–37; Johnson and Katcher, *Mines Against Japan*, p. 287.

18. Elliott, *Allied Minesweeping*, pp. 10, 25, 35.

19. *Conway's Fighting Ships, 1922–1946*, p. 87. Most PCEs were converted solely to escorts early in the war.

20. Furer, *Administration of the Navy Department*, pp. 107–15; Love, *Chiefs of Naval Operations*, p. xviii; McEathron, "Minecraft in the Van," pp. 2, 435.

21. Navy Department, U.S. Naval Administration in World War II, Bureau of Ordnance (hereafter BUORD Administrative History), "Underwater Ordnance," pp. 150–51; ONR, "Research and Development," pp. 122–27, 156–67; CNO, "Mine Warfare," pp. 35, 177; Duncan, *America's Use of Sea Mines*, p. 94; Lott, *Most Dangerous Sea*, pp. 18, 33. Fullinwider supervised the design and use of the Mk–6 mine in World War I and began the Mine Building that became NOL and the Mine Depot at Yorktown. Recalled from retirement a second time in 1940 by BUORD, he was also instrumental in founding the Mine Warfare School at Yorktown, the Mine and Bomb Disposal Schools around Washington, D.C., the Mine Warfare Test Station at Solomons, Maryland, and the Mine Warfare Operational Research Group. Hartmann, *Weapons That Wait*, pp. 51–52.

22. ONR, "Research and Development," pp. 172, 494–95, 500a; Rowland and Boyd, *Bureau of Ordnance*, p. 73; NOL Administrative History, part 1, pp. 17–35. For an explanation of the American degaussing experience, see Rowland and Boyd, *Bureau of Ordnance*, pp. 72–89.

23. Hartmann, "Mine Warfare History and Technology," pp. 22–23; *The Navy Department: A Brief History Until 1945* (Washington, 1970), p. 18; CNO, "Mine Warfare," pp. 12–29; BUORD Administrative History, p. 148.

24. CNO, "Mine Warfare," pp. 177–79; ONR, "Research and Development," p. 501; Lott, *Most Dangerous Sea*, pp. 66–67. The Solomons Island test station later became known as the Countermeasures Department. For a history of this facility, see BUORD Administrative History, Miscellaneous Activities, vol. 1, "A History of the U.S. Naval Mine Warfare Test Station, Solomons, Maryland."

25. ONR, "Research and Development," p. 501; Feldman, "Historical Review of Mine Countermeasures," p. 11; Lott, *Most Dangerous Sea*, pp. 33–34, 65; McEathron, "Minecraft in the Van," p. 435; Miller, "Minesweepers," p. 212. For one account of the early experiences of the converted fishing boats, see William Sheppard, "Dismal Spit and Her Mackerel Taxis," *Proceedings* 70 (Oct 1944): 1253–57; Elliott, *Allied Minesweeping*, p. 76.

26. *Mine Warfare Notes*, Mine Warfare Section, CNO, WW II Command File, OA; Johnson and Katcher, *Mines Against Japan*, pp. 44–45; McEathron, "Minecraft in the Van," p. 4; Rudy C. Garcia, "Yorktown, Va.: Mineman's Alma Mater," *All Hands* 481 (Feb 1957): 16, 18.

27. Rowland and Boyd, *Bureau of Ordnance*, p. 161; Johnson and Katcher, *Mines Against Japan*, p. 37; Hartmann, *Weapons That Wait*, pp. 127, 175; CNO, "Mine Warfare," pp. 66, 71–81; U.S. Strategic Bombing Survey, *The Offensive Mine Laying Campaign Against Japan* (1946; reprint,

Washington, 1969), pp. 26, 31. The early issues of *Mine Warfare Notes* were little more than carbon-copy newsletters. By the war's end, however, it had evolved into a polished professional journal. Some of the later editions featured mine warfare cartoons by artist Hank Ketcham, creator of "Dennis the Menace."

28. McEathron, "Minecraft in the Van," p. 440; Lott, *Most Dangerous Sea*, pp. 22–26; Duncan, *America's Use of Sea Mines*, pp. 94–95.

29. Michael Gannon, *Operation Drumbeat* (New York, 1990), p. 388; Lott, *Most Dangerous Sea*, p. 50; Ennis, "Deep, Cheap & Deadly," p. 27.

30. Hartmann, *Weapons That Wait*, p. 69; Elliott, *Allied Minesweeping*, p. 70; Howard A. Chatterton, "The Minesweeping/Fishing Vessel," *Proceedings* 96 (Jun 1970): 123; Jeffrey K. Bray, "Mine Awareness," *Proceedings* 113 (Apr 1987): 42–43; ONR, "Research and Development," pp. 998–99; Lott, *Most Dangerous Sea*, pp. 54–57.

31. Norman Friedman, *The Postwar Naval Revolution* (Annapolis, 1986), p. 178; ONR, "Research and Development," p. 503; McEathron, "Minecraft in the Van," p. 11. Building of these new minesweepers reached a peak for the Allied campaign in southern France in August 1944, and by the end of 1944 most had been transferred to the Pacific war.

32. Lott, *Most Dangerous Sea*, pp. 83–85. Samuel Eliot Morison, *Victory in the Pacific, 1945* (Boston, 1961), pp. 113–14; McEathron, "Minecraft in the Van," pp. 1, 37, Leopold, "Fleet Organization," pp. 14, 20. The command structure for the Pacific Fleet mine force is described in the 1958 MINEPAC Command History.

33. CNO, "Mine Warfare," pp. 99–101, 113–16; Johnson and Katcher, *Mines Against Japan*, p. 50.

34. McEathron, "Minecraft in the Van," p. 65–66; Samuel Eliot Morison, *Sicily–Salerno–Anzio, January 1943–June 1944* (Boston, 1962), pp. 175, 259, 273, 340–48; Elliott, *Allied Minesweeping*, pp. 84–86.

35. Elliott, *Allied Minesweeping*, p. 112; Hartmann, *Weapons That Wait*, pp. 70–71, 91; Samuel Eliot Morison, *The Invasion of France and Germany, 1944–1945* (Boston, 1957), pp. 46–47, 173.

36. Johnson and Katcher, *Mines Against Japan*, pp. 50, 61–62; Hartmann, *Weapons That Wait*, p. 131; Elliott, *Allied Minesweeping*, pp. 87–88. See G. K. Hartmann, *Wave Making by an Underwater Explosion*, NSWC/WOL MP 76–15, Naval Surface Weapons Center, White Oak, MD, 1976.

37. Morison, *The Invasion of France*, pp. 78–79; Elliott, *Allied Minesweeping*, pp. 107–8, 113–15, 126; Lott, *Most Dangerous Sea*, pp. 187–88. American vessels accounted for less than one-fourth of the minesweeping vessels in the American sector.

38. Elliott, *Allied Minesweeping*, pp. 118–20.

39. Lott, *Most Dangerous Sea*, p. 195; Hartmann, *Weapons That Wait*, p. 71; Morison, *Invasion of France*, pp. 216–17; Elliott, *Allied Minesweeping*, pp. 93, 129–31. Only divers could clear Katie mines.

40. CNO, "Mine Warfare," pp. 175–79.

41. Elliott, *Allied Minesweeping*, pp. 88–91; Johnson and Katcher, *Mines Against Japan*, pp. 60–62, 288; Hartmann, *Weapons That Wait*, p. 92; Feldman, "Historical Review of Mine Countermeasures," p. 15–16. For more detailed analysis of specific experimental sweep types, see NOL Administrative History, part 3, pp. 165–70.

42. Hartmann, *Weapons That Wait*, p. 135; Johnson and Katcher, *Mines Against Japan*, p. 289; Elliott, *Allied Minesweeping*, p. 148. For a history of the establishment of such ordnance disposal teams, see "U.S. Navy Mine and Bomb Disposal in World War II," Mine Warfare Section, CNO, WW II Command File, OA.

43. Elliott, *Allied Minesweeping*, p. 141; Morison, *Invasion of France*, pp. 286, 313; *Mine Warfare Notes*, 15 Jul 1945, p. 32.

44. Lott, *Most Dangerous Sea*, p. 32. The Japanese ultimately lost twenty-one merchant vessels to their own mines. Ibid., p. 82.

45. Hartmann, "Mine Warfare History and Technology," pp. 24–25; William L. Greer and James Bartholomew, "The Psychology of Mine Warfare," *Proceedings* 112 (Feb 1986): 58; U.S. Strategic Bombing Survey, pp. 2, 64. The U.S. first mined Haiphong with thirty-two Mk 12 mines with arming delays on 29 October 1942. British forces laid an additional forty U.S. mines at Haiphong between 16 October 1943 and 25 November 1944. Of the latter, the six Mk 13 Mod 5 mines reportedly accounted for three ships sunk in 1943. Total ship casualties at Haiphong throughout the war included five ships sunk and two damaged. U.S. Strategic Bombing Survey, pp. 64, 70, and part 1 of Appendix A.

46. McEathron, "Minecraft in the Van," p. 24; Elliott, *Allied Minesweeping*, pp. 7, 19, 40, 68–76; Morison, *Victory in the Pacific*, pp. 113–14; Hartmann, *Weapons That Wait*, pp. 136–38; ONR, "Research and Development," pp. 503–7.

47. *Conway's Fighting Ships, 1922–1946*, p. 150; "U.S. Navy Mine and Bomb Disposal in World War II," pp. 25, 41, 51. These ships, converted to AMCU for Underwater Locator in 1945, were never used operationally during the war, and will be discussed later.

48. Navy Department, "Administrative History of Minecraft," pp. 164–65; Samuel Eliot Morison, *Aleutians, Gilberts and Marshalls, June 1942–April 1944* (Boston, 1961), p. 158; Lott, *Most Dangerous Sea*, pp. 85, 168; Samuel Eliot Morison, *Leyte: June 1944–January 1945* (Boston, 1961), pp. 34–35, 118, 131; Morison, *Victory in the Pacific*, p. 113; MINEPAC Command History, 1958; Elliott, *Allied Minesweeping*, pp. 169, 180; McEathron, "Minecraft in the Van," p. 1, 37. Throughout the Pacific campaigns Rear Admiral Sharp exercised both administrative command of the Pacific Fleet minecraft through his staff in Pearl Harbor and personal command of the mine clearance operations in the Pacific. Lott, *Most Dangerous Sea*, p. 85.

49. Dudley W. Knox, *A History of the U.S. Navy* (New York: G. P. Putnam's Sons, 1948), p. 614; Elliott, *Allied Minesweeping*, pp. 158–67.

50. McEathron, "Minecraft in the Van," pp. 605–22.

51. Hartmann, *Weapons That Wait*, p. 77; CNO, "Mine Warfare," pp. 172–74, 185–90.

52. Rowland and Boyd, *Bureau of Ordnance*, p. 168; Duncan, *America's Use of Sea Mines*, pp. 122, 154–57; Johnson and Katcher, *Mines Against Japan*, p. 61; CNO, "Mine Warfare," p. 225.

53. Johnson and Katcher, *Mines Against Japan*, p. 33; Hartmann, *Weapons That Wait*, p. 196; Bertram Vogel, "The Great Strangling of Japan," *Proceedings* 73 (Nov 1947): 1307; U.S. Strategic Bombing Survey, pp. 1–3. In October 1945 Commander Seburo Tadenuma, IJN, described the mines as "one of the main causes of our defeat," whereas Captain Kyuzo Tamura, IJN, claimed that "the result of the B–29 mining was so effective against the shipping that it eventually starved the country. I think you probably could have shortened the war by beginning earlier." U.S. Strategic Bombing Survey, p. 3.

54. The Minecraft, Pacific Fleet, organization was still incomplete when demobilization began in August 1945. Most of the new ships and staffers had not yet arrived, and no satisfactory flagship had been found for Rear Admiral Sharp. Of the 506 ships under his command, 412 were minesweepers and hunters. Another 57 had not yet arrived, and the remainder were minelayers and auxiliary vessels. Sixty-four percent of his MCM vessels were YMSs. Navy Department, "Administrative History of Minecraft," p. 6.

55. James E. Auer, *The Postwar Rearmament of Japanese Maritime Forces, 1945–71* (New York, 1973), pp. 49–52.

56. Malcom W. Cagle and Frank A. Manson, *The Sea War in Korea* (Annapolis, 1957), p. 137; Charles G. McIlwraith Comment, *Proceedings* 100 (Jul 1974): 88; John D. Alden, "The

Indestructible XMAP," *Naval History* 2 (Winter 1988): 45; Elliott, *Allied Minesweeping*, pp. 170–74, 175–78; CNO, "Mine Warfare," p. 191. Figures for mines swept are from *Mine Warfare Notes*, 15 Oct 1945, p. 12, and include all mines cleared between 7 December 1941 through 15 October 1945. A. D. Van Nostrand, "Minesweeping," *Proceedings* 72 (Apr 1946): 505. The quote is from the 1958 MINEPAC Command History. Ironically, paravanes had been largely abandoned during the war because of the threat of influence mines, which few ships ultimately encountered. Proliferation of large bow sonar domes also contributed to their permanent demise. Norman Friedman, *World Naval Weapons Systems* (Annapolis, 1989), p. 445.

57. *New York Times*, 15 Jan 1947; Auer, *Postwar Rearmament*, pp. 49–52. Because of continuing reappearance of sea mines throughout the Pacific, even today most ships still keep to known swept channels.

58. Elliott, *Allied Minesweeping*, pp. 188–89, 190–91. Numbers of ships include AMs, AMCs, trawlers, YMSs, and PCSs briefly used, but not experimental egg crates. Of the casualties from all causes, 26 were fleet minesweepers and 32 were others, mostly YMSs. Of the latter, 11 were sunk by mines. Ibid., pp. 78–79, 190–91. In 1948, King estimated that a force of 874 MCM vessels had been added to the fleet. Ernest J. King, *U.S. Navy at War, 1941–1945. Official Reports to the Secretary of the Navy by Fleet Admiral Ernest J. King, USN* (Washington, 1946), p. 284.

59. Michael A. Palmer, *Origins of the Maritime Strategy: American Naval Strategy in the First Postwar Decade* (Washington, 1988): 7, 13, 22–27, 62; Samuel Eliot Morison, *Liberation of the Philippines: Luzon, Mindanao, the Visayas, 1944–1945* (Boston, 1969), pp. 268–70; Auer, *Postwar Rearmament*, p. 63.

60. MINEPAC Command Histories, 1958, 1959; MINELANT Command Histories, 1959, 1966, OA. A "mine force commander" existed under the Service Force, Atlantic Fleet, from December 1945 until creation of the type command in April 1946. In the 1958 MINEPAC Command History, the writer described this command as "dissolved" in January 1947 and recorded its "reestablishment" in January 1951.

61. U.S. Strategic Bombing Survey, pp. 30–32.

3. The Wonsan Generation

1. Elliott, *Allied Minesweeping*, p. 7; Van Nostrand, "Minesweeping," p. 508; CNO, "Mine Warfare," p. 229.

2. ONR, "Research and Development," pp. 502–3; Naval Coastal Systems Center, 1984 Command History File, OA. The laboratory (renamed the U.S. Navy Mine Defense Laboratory in 1955, the Naval Ship Research and Development Laboratory in 1968, the Naval Coastal Systems Laboratory in 1972, and finally the Naval Coastal Systems Center in 1978) was responsible for all undersea countermeasures, special warfare, amphibious warfare, diving and salvage research, development, and testing.

3. Miller, "Minesweepers," p. 212; *Conway's All the World's Fighting Ships, 1947–1982*, Part 1, *The Western Powers* (Annapolis, 1983), pp. 191, 253; Cagle and Manson, *The Sea War*, p. 127; Hartmann, *Weapons That Wait*, pp. 89, 135, 232; Lott, *Most Dangerous Sea*, pp. 70–71; Philip Nepean, "Naval Mine Countermeasures," *Armada International* 8 (Mar 1984): 106.

4. McEathron, "Minecraft in the Van," p. 624; Edwin Bickford Hooper, Dean C. Allard, and Oscar P. Fitzgerald, *United States Naval Operations in the Vietnam Conflict*, vol. 1, *The Setting of the Stage to 1959* (Washington, 1976), pp. 104–9; USN Press Release No. 127–46, 23 Mar 1946, VADM Arthur D. Struble Biographical File, OA; *Conway's Fighting Ships, 1947–1982*, 1:191.

5. Palmer, *Maritime Strategy*, p. 66; Edward J. Marolda and G. Wesley Pryce III, *A Short History of the United States Navy and the Southeast Asian Conflict, 1950–1975* (Washington, 1984), p. 1; William B. Fulton, *Riverine Operations, 1966–1969* (Washington, 1973), pp. 13–14. At least three

coastal minesweepers (MSC), the design derivative of the YMS, joined the French Navy for minesweeping in Vietnam. R. L. Schreadley, "The Naval War in Vietnam, 1950–1970," *Proceedings, Naval Review 1971* 97 (May 1971), pp. 182–84.

6. "Mine Countermeasures: Present Situation and Necessary Future Action," enclosure to CNO to Chiefs of the Office of Naval Research, Bureau of Ordnance, Bureau of Ships and Chairman, Navy Research and Development Review Board, ser 00018P31, 17 Apr 1950, box 8, Project I–50, General Board Studies 1949–50, RG 80, DNA (hereafter CNO Mine Countermeasures Study, 1950).

7. F. S. Low, "A Study of Undersea Warfare," 22 Apr 1950, pp. 3, 21, 84, 109–110 (hereafter "Low Report") Command File, OA.

8. Massachusetts Institute of Technology, "Project Hartwell: A Report on the Security of Overseas Transport," 2 vols., 21 Sep 1950, 1: F–13, F–2, Post-1 Jan 1946 Command File, OA.

9. James A. Field, Jr., *History of United States Naval Operations: Korea* (Washington, 1962), pp. 45–46, 54, 231. The YMSs were redesignated AMS in 1947.

10. Bryan Ranft and Geoffrey Till, *The Sea in Soviet Strategy* (Annapolis, 1983), pp. 90–91, 96; Norman Polmar, *Soviet Naval Power: Challenge for the 1970s* (New York, 1974), pp. 3–30; John Chomeau, "Soviet Mine Warfare," *Naval War College Review* 24 (Dec 1971): 94–96.

11. Edwards, "A Naval Lesson of the Korean Conflict," p. 1339; Anthony Preston, "The Infernal Machine: Mines and Mine Countermeasures," *Defence* 19 (Aug 1988): 561; Field, *Korea*, p. 183; Malcolm W. Cagle and Frank A. Manson, "Wonsan: the Battle of the Mines," *Proceedings* 83 (Jun 1957): 602; Cagle and Manson, *The Sea War*, p. 129.

12. Field, *Korea*, pp. 193, 217, 231–32. Ships damaged included *Brush* off Tanchon, *YMS–509* off Yongdok, *Mansfield* off Changjon, and *YMS–504* off Mokpo. *Magpie* lost two thirds of her crew when the contact mine caught up in her sweep gear came loose and hit her. Cagle and Manson, "Wonsan," p. 603.

13. Field, *Korea*, pp. 220, 230; Edwards, "A Naval Lesson of the Korean Conflict," p. 1337. Lacking sufficient minesweeping vessels, General MacArthur was convinced to attempt the Wonsan landing rather than one at Hungnam. Cagle and Manson, "Wonsan," pp. 603–4.

14. Struble Biographical File, CAPT Richard T. Spofford Biographical File, OA. Spofford's mine warfare experience included Naval Postgraduate School training in "Ordnance Engineering (Mines)," 1936; mine officer on staff of Commander Battle Force, Pacific, with additional duty on staff of Commander Service Squadron (COMSERVRON) 6, 1939–1942; OPNAV mine warfare specialist, 1942–1943; weapons officer, Mine Warfare Test Station, Solomons, Maryland, 1943–1944; Commander Mine Squadron (COMINRON) One, 1945; commanding officer of minelayer *Terror* (CM–5), 1945–1947; and ordnance inspector, 1947–1950. After serving as COMINRON 3 and Commander Task Group (CTG) 95.6 from August 1950 to March 1951, he served as plans officer, Minecraft, Pacific Fleet, 1951–1953; and as head of Harbor Defense Branch and assistant director of OPNAV Undersea Warfare Division. Spofford described his specialty as "field testing of new service mines and homing torpedoes."

15. Cagle and Manson, *The Sea War*, pp. 118–20. E. B. Potter in *Admiral Arleigh Burke* (New York, 1990), p. 343, claims that Joy and Burke opposed the Wonsan invasion because of suspected mining, but that MacArthur "refused even to discuss any change" in plans.

16. Auer, *Postwar Rearmament*, pp. 49–52, 64–65; Potter, *Burke*, pp. 343–44.

17. Cagle and Manson, *The Sea War*, pp. 133–35; Cagle and Manson, "Wonsan," pp. 604–5; Field, *Korea*, pp. 222–26, 232–33. Task Group 95.6 consisted of twenty-four ships: 1 DD, 1 APD, 2 DMSs, 3 AMs, 7 AMSs, 1 ARG, 1 ARS, and 8 Japanese JMSs.

18. Cagle and Manson, *The Sea War*, p. 135; Cagle and Manson, "Wonsan," p. 605; Field, *Korea*, pp. 222–26, 232–33.

19. LT C. E. McMullen to SECNAV, ser 01A–50, 19 Oct 1950, *Pirate* file, Ships' Histories Branch, Naval Historical Center (hereafter SH); Edwin H. Simmons, "Mining at Wonsan–and in the Persian Gulf," *Fortitudine* 17 (Summer 1987): p. 6.

20. Cagle and Manson, *The Sea War*, pp. 136–42; Field, *Korea*, pp. 233–37; Cagle and Manson, "Wonsan," pp. 606–9; Edwards, "A Naval Lesson of the Korean Conflict," pp. 1337–38; C. J. Wages, "Mines . . . The Weapons That Wait," *Proceedings* 88 (May 1962): 103.

21. Auer, *Postwar Rearmament*, p. 65; Preston, "The Infernal Machine," p. 561; Roy F. Hoffman, "Offensive Mine Warfare: A Forgotten Strategy?" *Proceedings, Naval Review* 103 (May 1977): 150; Cagle and Manson, *The Sea War*, pp. 136–44; Field, *Korea*, pp. 233–37; Cagle and Manson, "Wonsan," pp. 598, 604, 606–9; Edwards, "A Naval Lesson of the Korean Conflict," pp. 1337–38.

22. Paolo E. Coletta, "Naval Mine Warfare," *Proceedings* 85 (Nov 1959): 84; Max Hastings, *The Korean War* (New York, 1987), pp. 125–26; Simmons, "Mining at Wonsan," p. 4; Edwards, "A Naval Lesson of the Korean Conflict," p. 1337; *Newsweek*, 7 Nov 1950, p. 30.

23. Auer, *Postwar Rearmament*, p. 65.

24. Mine Advisory Committee, National Academy of Sciences—National Research Council, "PROJECT NIMROD: The Present and Future Role of the Mine in Naval Warfare" (hereafter "Project Nimrod"), 15 Jan 1970, Chapter 2 (unclassified), p. 67.

25. Cagle and Manson, *The Sea War*, p. 151; Cagle and Manson, "Wonsan," p. 611.

26. Cagle and Manson, *The Sea War*, p. 142.

27. Field, *Korea*, pp. 237–42, 246; Cagle and Manson, *The Sea War*, p. 219; "Project Nimrod," p. 68; Feldman, "Historical Review of Mine Countermeasures," p. 14–15.

28. Cagle and Manson, *The Sea War*, pp. 193–94, 198–202; Field, *Korea*, pp. 249, 319, 331, 345. *LST-799* served as MCM and helicopter mother ship from 1951.

29. Report of the Secretary of the Navy, 30 Jul 1951, U.S. Department of Defense, *Semiannual Report of the Secretary of Defense and the Semiannual Reports of the Secretary of the Army, Secretary of the Navy, Secretary of the Air Force, January 1 to June 30 1952* (Washington, 1952) (hereafter *Semiannual Report of the Secretary of Defense*), p. 189; Feldman, "Historical Review of Mine Countermeasures," p. 7; Naval Sea Systems Command, "Operation 'End Sweep': A History of Mine Sweeping Operations in North Vietnam" (Sanitized) (hereafter NAVSEA, "End Sweep"), p. 2–2; Cagle and Manson, *The Sea War*, pp. 214–18; Cagle and Manson, "Wonsan," p. 610; Field, *Korea*, pp. 358–59, 447; CINCPACFLT, *Korean War, U.S. Pacific Fleet Operations, Mine Warfare, Interim Evaluation Reports*, p. 8-1, OA; Auer, *Postwar Rearmament*, p. 66.

30. Hartmann, "Mine Warfare History and Technology," p. 27; Norman Friedman, "Postwar British Mine Countermeasures–And National Strategy," *Warship* 41 (Jan 1987): 46; Cagle and Manson, *The Sea War*, pp. 220–21; Edwards, "A Naval Lesson of the Korean Conflict," p. 1338.

31. MINEPAC Command Histories, 1958 and 1970, OA; Field, *Korea*, p. 242; Cagle and Manson, *The Sea War*, pp. 220–21; R. L. Schreadley, "The Mine Force–Where the Fleet's Going, It's Been," *Proceedings* 100 (Sep 1974): 27. The previous name of the Pacific Fleet type command had been Minecraft, Pacific Fleet (1944–1947). In the 1970 MINEPAC Command History that writer records the "reactivation" of MINEPAC in 1951. All MINEPAC listings of command organization for all years in all command histories leave a gap in MINEPAC leadership from the disestablishment of Minecraft, Pacific Fleet, in January 1947 to the establishment of Mine Force, Pacific Fleet (also known as MINEPAC), in January 1951.

32. W. M. Emshwiller Comment, *Proceedings* 101 (Feb 1975): 76; *Implicit* (MSO 455) Command History, 1965, SH; *Prime* Command History File, SH; Philip L. Rhodes obituary, *New York Times*, 30 Aug 1974. Of a total of 101 MSOs completed between 1953 and 1960, 65 were retained by the U.S. Navy, and 36 were transferred to foreign countries. John S. Roye and Samuel L. Morison, *Ships and Aircraft of the U.S. Fleet* (Annapolis, 1972), p. 88.

33. *Conway's All the World's Fighting Ships, 1947–1982*, 1:191, 252.

34. Naval History Division, *Dictionary of American Naval Fighting Ships* (Washington, 1970), 5:442, 481; Friedman, *Postwar Naval Revolution*, p. 183.

35. Cagle and Manson, *The Sea War*, pp. 135, 197, 218–19; Paul L. Gruendl, "U.S. Navy Airborne Mine Countermeasures: A Coming of Age," Professional Study No. 5617, Air War College, Maxwell Air Force Base, AL, 1975 (hereafter "USN AMCM"), p. 34.

36. Gruendl, "USN AMCM," pp. 16–19; John A. H. Torry, "Minesweeping From the Air," *Proceedings* 87 (Jan 1961): 139–40; Mort Schultz, "They Hunt for Floating Death," *Popular Mechanics* 129 (Apr 1968): 214; Schreadley, "The Mine Force," p. 31.

37. H. George Baker, "The MSB Story– Little Ships Sweep the Sea to Keep it Free," *All Hands* 481 (Feb 1957): 2–5; Miller, "Minesweepers," p. 219; Norman Friedman, "United States of America," in *Conway's Fighting Ships, 1947–1982*, 1:191–92, 200, 251–53; Coletta, "Naval Mine Warfare," p. 96. Many more MSCs and several MSOs were also built for other navies.

38. Hartmann, *Weapons That Wait*, pp. 80–81, 127, 133, 136; "Project Nimrod," pp. 68–69; J. B. Chaplin, "The Application of Air Cushion Technology to Mine Countermeasures in the United States of America," *International Symposium on Mine Warfare Vessels and Systems, London, 12–15 June 1984* (London, 1984), 2:2; NAVSEA, "End Sweep," p. 2–12. The Mine Advisory Committee was later incorporated into the Naval Studies Board.

39. John Alden, "The Indestructible XMAP," pp. 45–46; Miller, "Minesweepers," p. 219. After she foundered off the Florida coast in blast testing in 1961, XMAP was salvaged in 1964 and sold for scrap. Alden, "The Indestructible XMAP," p. 47.

40. *Conway's Fighting Ships, 1947–1982*, 1:191, 253–54; Alden, "The Indestructible XMAP," p. 45; R. L. Schreadley Comment, *Proceedings* 101 (Dec 1975): 81. These Liberty Ships were reclassified MSS (Minesweeper, Special). LST–1166, reclassified MSS–2, was used to check sweep in Operation Endsweep in 1973.

41. Stanley J. Norman, "Prepared for Mine Warfare?" *Proceedings* 109 (Feb 1983): 67; Alfred T. Hamilton, "Clearing the Way: LANTFLT Mine Ops," *Surface Warfare* 6 (Sep 1981): 30; "Mine Countermeasures," p. 282. SQQ–14's descendants, SQQ–30 and 32, are currently being placed on MCM–1 vessels.

42. [Hooper], *The Navy Department*, p. 40; "Project Nimrod," p. 69; Hartmann, *Weapons That Wait*, pp. 260–61; Friedman, "US Mine-Countermeasures," p. 1259.

43. H. George Baker, "The Mine Force: Wooden Ships and Iron Men," *All Hands* 481 (Feb 1957): 6; Report of the Secretary of the Navy, 22 Sep 1952, *Semiannual Report of the Secretary of Defense*, pp. 159–60; Robert Winters, "History and Analysis: Mine Warfare Operations in Southeast Asia, 1961–1973," Post-1 Jan 1946 Command File, OA; "Port Protectors," *All Hands* 481 (Feb 1957): 21.

44. Baker, "The Mine Force: Wooden Ships and Iron Men," pp. 6–9; Friedman, *Postwar Naval Revolution*, pp. 176–77; *Dictionary of American Naval Fighting Ships*, 5:444; *Conway's Fighting Ships, 1947–1982*, 1:191–92, 200, 251, 252, 253; Schreadley, "The Mine Force," p. 28; Miller, "Minesweepers," p. 212. In 1972 the Mine Warfare School combined with the Fleet Training Center at Charleston to become the Fleet and Mine Warfare Training Center.

45. VADM Joseph Donnell, Commander Naval Surface Force, Atlantic Fleet (former CO *Kingbird* [MSC–194], 1958–1960), interview with author, Sep 1988, Norfolk, VA; CAPT (RADM-S) Robert C. Jones, OP–09B (former XO *Dominant* [MSO–431], 1967–1968), interview with author, 24 May 1989, Pentagon, Washington, DC.

46. *Conway's Fighting Ships, 1947–1982*, 1:251.

47. "Project Nimrod," p. 70.

4. New Lessons Learned

1. Marolda and Pryce, *Short History*, pp. 1, 4; Schreadley, "Naval War in Vietnam, pp. 182–84.

2. James A. Hoffman, "Market Time in the Gulf of Thailand," *Naval Review 1968* (Annapolis, 1968): 47; Marolda and Pryce, *Short History*, pp. 25, 46; Edwin Bickford Hooper, *Mobility, Support, Endurance: A Story of Naval Operational Logistics in the Vietnam War, 1965–1968* (Washington, 1972), pp. 129–30.

3. Marolda and Pryce, *Short History*, pp. 7–8, 12–13; Schreadley, "Naval War in Vietnam," p. 185; Hooper, *Mobility, Support, Endurance*, p. 13; Murland W. Searight, "Prepare to Sweep Mines. . . ," *Proceedings* 96 (Jan 1970): 55–59.

4. William B. Fulton, *Riverine Operations, 1966–1969* (Washington, 1973), pp. 3, 96–97; Carl White, "Move and Countermove: Belated Recognition for Naval Mine Warfare and Mine Countermeasures Requirements," *Sea Power* 28 (Jun 1985): 12; Thomas J. Cutler, *Brown Water, Black Berets: Coastal and Riverine Warfare in Vietnam* (Annapolis, 1988), p. 47; Mort Schultz, "They Hunt For Floating Death," *Popular Mechanics* 129 (Apr 1968): 86; S. A. Swarztrauber, "River Patrol Relearned," *Proceedings* 96 (May 1970): 122.

5. Hooper et al., *The Setting of the Stage*, p. 228; William H. Cracknell, Jr, "The Role of the U.S. Navy in Inshore Waters," *Naval War College Review* 21 (Nov 1968): 75; Cutler, *Brown Water, Black Berets*, pp. 182–83; MINEPAC Command History, 1966, OA.

6. Cracknell, "The Role of the U.S. Navy in Inshore Waters," p. 80; Swarztrauber, "River Patrol Relearned," pp. 123–28; Victor Croziat, *The Brown Water Navy: The River and Coastal War in Indo-China and Vietnam, 1948–1972* (Dorset, UK, 1984), p. 156.

7. Clark M. Gammell, "Naval and Maritime Events, 1 July 1966–30 June 1967," *Naval Review 1968* (Annapolis, 1968): 257. After two years of river duty and the receipt of the Presidential Unit Citation, detachment Alfa was redesignated Mine Division 112 and given six MSBs and five MSMs. Mine Division 113, operating in the Rung Sat, was similarly manned. Swartztrauber, "River Patrol Relearned," pp. 142–48.

8. Robert E. Mumford, Jr., "Jackstay: New Dimensions in Amphibious Warfare," *Naval Review 1968* (Annapolis, 1968): 74; Marolda and Pryce, *Short History*, pp. 52–53; Swarztrauber, "River Patrol Relearned," pp. 125, 142; George R. Kolbenschlag, "Minesweeping on the Long Tao River," *Proceedings* 93 (Jun 1967): 90–94; Fulton, *Riverine Operations*, p. 175; Hooper, *Mobility, Support, Endurance*, pp. 164–65; Schreadley, "Naval War in Vietnam," p. 204.

9. Schreadley, "Naval War in Vietnam," pp. 194–95, 201–4; Swarztrauber, "River Patrol Relearned," pp. 128, 142, 151, 156; Hooper, *Mobility, Support, Endurance*, p. 168. For more detailed information on mine and guerrilla attacks on U.S. riverine forces, see U.S. Naval Forces, Vietnam, Monthly Historical Summaries (monthly, Apr 1966–Dec 1971; quarterly, Jan 1972–Mar 1973), OA.

10. Hooper, *Mobility, Support, Endurance*, pp. 122–23; Swarztrauber, "River Patrol Relearned," pp. 92–93; 129–33, 142–43; Marolda and Pryce, *Short History*, p. 55.

11. Croziat, *The Brown Water Navy*, p. 149; *Conway's Fighting Ships, 1947–1982*, 1:192, 254; NAVSEA, "End Sweep," pp. 1-3 to 1-6, 3-19, A-13, A-17.

12. Marolda and Fitzgerald, *From Military Assistance to Combat, 1959–1965*, p. 317; Schultz, "They Hunt for Floating Death," p. 89; Kolbenschlag, "Minesweeping on the Long Tao River," p. 98; "Radio-Controlled Drone Boats Used in Vietnam Minesweeping," *Proceedings* 96 (Feb 1970): 123–24; Hartmann, *Weapons That Wait*, p. 209.

13. Friedman, *World Naval Weapons Systems*, p. 476; Norman Polmar, *Ships and Aircraft of the U.S. Fleet* (Annapolis, 1985), p. 233.

14. Croziat, *The Brown Water Navy*, p. 156; *Conway's Fighting Ships, 1947–1982*, 1:186.

15. John Van Nortwick, "Endsweep," *Marine Corps Gazette* 58 (May 1974): 36.

16. Gruendl, "USN AMCM," pp. 20–26.

17. James M. McCoy, "Mine Countermeasures: Who's Fooling Whom?" *Proceedings* 101 (Jul 1975): 40; Schreadley, "The Mine Force," pp. 29–31; J. B. Bonds Comment, *Proceedings* 101 (Jun 1975): 78; Beaver, "Aerial Minesweeping," pp. 306–7; *Conway's Fighting Ships, 1947–1982*, 1:191; Emshwiller Comment, *Proceedings* 101 (Jun 1975): 76; Hartmann, *Weapons That Wait*, pp. 132, 139; Schultz, "They Hunt for Floating Death," p. 214; Schreadley, "The Mine Force," p. 31; Torry, "Minesweeping From the Air," pp. 139–40; NAVSEA, "End Sweep," p. vii; MINRON 8 Command History, 1970, OA.

18. Quoted in Norman Polmar, "The U.S. Navy: Mine Countermeasures," *Proceedings* 105 (Feb 1979): 117.

19. Elmo R. Zumwalt, Jr., *On Watch: A Memoir* (New York, 1976), pp. 3–4, 72–73, 80–81; Gruendl, "USN AMCM," pp. 37–39; NAVSEA, "End Sweep," p. 2-2. Some of the wooden ships went to the Naval Reserve Force, which kept them active in reserve on both coasts. Schreadley, "The Mine Force," pp. 27–29; Gary P. Schwartzkopf Comment, *Proceedings* 101 (Jan 1975): 78; MINELANT Command History, 1970, OA.

20. Truver, "Weapons That Wait," pp. 39–40; Hartmann, *Weapons That Wait*, p. 147; Zumwalt, *On Watch*, p. 81; NAVSEA, "End Sweep," p. 2-2.

21. Gruendl, "USN AMCM," pp. 38–40; Winters, "History and Analysis," OA. The 1971–1973 MCM shipbuilding program included sixteen new design MSOs and two new design MSCs. Winters, "History and Analysis," OA.

22. Gruendl, "USN AMCM," pp. 41–43; Hooper, *Mobility, Support, Endurance*, p. 32. The quote is from *Parrot* (MSC–197) Command History, 1972, SH. Prior to the creation of MINEWARFOR, Mine Forces, Atlantic and Pacific, retained only administrative control of their ships, although ships often reported that the type commander also "maintained operational control for the greater part of the year." *Shrike* (MSC–201) Command History, 1968, SH.

23. Gruendl, "USN AMCM," pp. 43, 47–49; MOMCOM Command History, 1972, OA.

24. Gruendl, "USN AMCM," pp. 28–29, 52–54.

25. Marolda and Pryce, *Short History*, pp. 86–89; William Paden Mack, "As I Recall. . . ," *Proceedings* 106 (Aug 1980): 105; NAVSEA, "End Sweep," p. 1-3. The quote is from Brian McCauley, "Operation End Sweep," *Proceedings* 100 (Mar 1974): 19. MINEWARFOR planners included Commander Paul L. Gruendl and Lieutenant James M. McCoy. COMINEWARFOR planned for CINCPACFLT, COMSEVENTHFLT, and the carrier division, as well as for the squadron executing the operation.

26. Mack, "As I Recall," p. 105; Marolda and Pryce, *Short History*, pp. 86–89; Greer and Bartholomew, "The Psychology of Mine Warfare," p. 60; William Paden Mack, "As I Recall," *Proceedings* 106 (Aug 1980): 105; NAVSEA, "End Sweep," p. 4-2.

27. NAVSEA, "End Sweep," p. 3-31; MINEWARFOR Command History, 1972, OA.

28. Gruendl, "USN AMCM," p. 29. The original name for the mine countermeasures plan, Formation Sentry, became Formation Sentry II when clearance requirements widened during negotiations, and Marine Corps helicopters became available. Winters, "History and Analysis," OA. The Swept Mine Locator System used gyrocompass bearings and radar or stadimeter ranges.

29. CTF 78 to CTF 77, "Mine Warfare Lessons Learned," 3 Jul 1973, PL 08 C, p. 2, COMINEWARCOM Archives, Fleet and Mine Warfare Training Center Library, Charleston, SC (hereafter McCauley, "Lessons Learned"); NAVSEA, "End Sweep," pp. 2-4 to 2-5; Gruendl, "USN AMCM," pp. 55–59.

30. Gruendl, "USN AMCM," pp. 62–63; NAVSEA, "End Sweep," pp. 3-2, A-13; Friedman, *World Naval Weapons Systems*, p. 477. Thirteen Navy and twenty-four reconfigured Marine Corps

CH–53Ds ultimately made up the task group complement. In addition to HMH–463 and HMM–165, CH–53 Sea Stallion helicopters were transferred from HMM–462, and CH–46 Sea Knight helicopters and crews from HMM–164 flew in logistic support. For more detail on these units, see Curtis G. Arnold and Charles D. Melson, *U.S. Marines in Vietnam: The War That Would Not End, 1971–1973* (Washington, forthcoming).

31. Marolda and Pryce, *Short History*, pp. 90–91; Hartmann, "Mine Warfare History and Technology," p. 27; Gruendl, "USN AMCM," pp. 63, 85, appendixes f and g; NAVSEA, "End Sweep," pp. 2-6, 2-8, 5-48.

32. Walter Scott Dillard, *Sixty Days to Peace: Implementing the Paris Peace Accords* (Washington, 1982), pp. 85–86; NAVSEA, "End Sweep," pp. 5–11, 5-19–5-20, 5-32–5-35, A-23; Marolda and Pryce, *Short History*, p. 93. This one and only mine detonation during End Sweep has caused considerable comment over the years. According to all surviving officers interviewed, the explosion really was a live mine, captured only on film because of the Swept Mine Locator System. Its detonation surprised everyone involved because they swept only areas where mines were believed to have gone inert. Other unverified mine explosions may indeed have been ordnance lobbed in the water by sailors interested in confusing the North Vietnamese.

33. Roy F. Hoffman. "Offensive Mine Warfare: A Forgotten Strategy?" *Proceedings* 103 (May 1977): 154; NAVSEA, "End Sweep," pp. 5-53–54, 5-37, 5-39, 5-43, 5-48; Schreadley, "The Mine Force," p. 31; McCoy, "Mine Countermeasures," p. 42; McCauley, "Operation End Sweep," p. 22. AMCM unit Delta completed the sweep on 5 July at Cua Sot. NAVSEA, "End Sweep," p. x.

34. Hartmann, *Weapons That Wait*, pp. 154, 271; McCauley, "Lessons Learned," Problems MA 02, MA 03, PL 06 C.

35. Gruendl, "USN AMCM," p. 70; John Van Nortwick, "Endsweep," *Marine Corps Gazette* 58 (May 1974): 31; Winters, "History and Analysis," OA.

36. McCauley, "Operation End Sweep," p. 23.

37. Ibid., pp. 24–25.

38. McCauley, "Lessons Learned," Problems PL 02 C, PL 05 C, PL 09 C, OP 04, and OP 06.

39. Ibid.; McCauley, "Lessons Learned," Problem PL 03 C.

40. NAVSEA, "End Sweep," pp. 6-2, A-7–9; John G. Robinson, "Miners Served Here," *Surface Warfare* 5 (Aug 1980): 27.

41. ADM I. C. Kidd to CINCPACFLT, Feb 1973, quoted in NAVSEA, "End Sweep," p. 3-4.

42. Gruendl, "USN AMCM," pp. 66, 100; Ennis, "Deep, Cheap & Deadly," p. 30.

43. Gruendl, "USN AMCM," p. 106; Larry L. Booda, "Mine Warfare Moves Ahead, Plays ASW Role," *Sea Technology* 25 (Nov 1984): 13; J. Huntly Boyd, Jr., "Nimrod Spar: Clearing the Suez Canal," *Proceedings* 102 (Feb 1976): 20; Navy Department, Naval Inshore Warfare Command, Atlantic, "Final Report, Suez Canal Clearance Operation, Task Force 65," May 1975 (hereafter "Final Report, TF 65"), pp. iii–v, II-5–7, II-30–31, VI-2.

44. The MCM and MHC Ship Acquisition Program Office (PMS–303) in the Naval Sea Systems Command is now responsible for designing, building, and delivering mine warfare ships, and reports to the platform director for amphibious, auxiliary, mine and sealift ships, who in turn reports to Commander NAVSEA. The Mine Warfare Systems Program Office (PMS–407) is responsible for designing, developing, and procuring mines and MCM systems (minehunting sonar, ROVs, precise navigation systems, shipboard command and control systems, wire sweeps, etc.) and reports to the deputy commander for weapons and combat systems who in turn reports to Commander NAVSEA. Supporting laboratories and contractors for these two program offices are divided into two camps, a workable but less efficient system that is the result of the bureaus' historical divisions. A similar situation once existed in the Aegis program management offices, but consolidation of those entities enhanced Aegis program effectiveness.

45. Naval Surface Force, U.S. Atlantic Fleet (SURFLANT) was established 1 January 1975, whereas Naval Surface Force, U.S. Pacific Fleet (SURFPAC) began on 1 April 1975. Mine warfare ships were transferred to each on 1 July 1975. SURFLANT and SURFPAC Command Histories, 1975, OA. The second and last Commander Mine Warfare Force was Rear Admiral Roy F. Hoffman (November 1974–July 1975) who served on two MCM vessels and survived the mining of *Pirate* (AM–275), sunk while sweeping at Wonsan.

46. OPNAVINST 3370.3B, 9 Jul 1982; Kenneth A. Heine, "Sweeping Ahead," *Surface Warfare* 13 (Mar-Apr 1988): 7; Remarks by CNO Admiral Carlisle A. H. Trost, COMINEWARCOM change of command, 8 Jul 1989. When COMINEWARCOM is dual-hatted as Commander Naval Base, Charleston, he is most often a two-star with dual responsibilities.

47. "Mine Countermeasures," *Navy International* 93 (Jun 1988): 280–82; Preston, "The Infernal Machine," pp. 559–61; Hessman, "In Search of 'The Weapon That Waits'," p. 41; Jeremy D. Taylor, "Mining: A Well Reasoned and Circumspect Defense," *Proceedings* 103 (Nov 1977): 44; James D. Hessman, "Mine Warfare: A Two-Sided Game," *Sea Power* 27 (Aug 1984): 30; Marriott, "A Survey of Modern Mine Warfare," p. 40; Hartmann, "Mine Warfare History and Technology," p. 27.

48. Norman Friedman, "US Mine-Countermeasures Programs," *International Defense Review* 17 (1984): 1260–64; Truver, "Weapons That Wait," p. 39; Polmar, "The U.S. Navy: Mine Countermeasures," p. 117; *Conway's All the World's Fighting Ships, 1947–1982*, 1:251.

49. Cyrus R. Christensen, "A Minesweeping Shrimp Boat? A What?" *Proceedings* 107 (Jul 1981): 110–11. MSSB-1, which became the prototype for the COOP program—named for the plan to use boats, often captured drug runners, that may opportunely become available—allowed modular navigational, sweep, and remote vehicle units to be placed on a variety of vessels. Naval reservists man the twenty-two planned vessels, conducting underwater surveys of harbors in peacetime and providing the nucleus of a harbor sweep fleet in wartime. Calling their program "ships taken up from trade" or "STUFT," the British again recalled their fishing trawlers to clear Argentine mines in the 1982 Falklands War. Preston, "The Infernal Machine," p. 564; R. Bell and R. Abel, "Mine Warfare CO-OPeration," *Proceedings* 110 (Oct 1984): 147–49; Michael S. Ruth, "COOP-The Breakout Gang," *Surface Warfare* 9 (Jul-Aug 1984): 19–20; USNI Seminar, "Mine Warfare"; Wesley McDonald, "Mine Warfare: A Pillar of Maritime Strategy," *Proceedings* 111 (Oct 1985): 56.

50. Ibid., p. 109; D. W. Cockfield, "Breakout: A Key Role For the Naval Reserve," *Proceedings* 111 (Oct 1985): 54; C. F. Horne, "New Role for Mine Warfare," *Proceedings* 108 (Nov 1982): 37; Friedman, "US Mine-Countermeasures Programs," p. 1259.

51. Hessman, "Mine Warfare: A Sweeping Assessment," p. 10; Curtis, "Mines at Sea," p. 22; Truver, "Weapons That Wait," p. 37; Friedman, "US Mine-Countermeasures Programs," p. 1259; "Mine Warfare: The U.S. Navy Reacts," *Navy International* 90 (Jan 1985): 35; United States Naval Institute Professional Seminar Series (hereafter USNI Seminar), "Mine Warfare: Which Platform?" 26 Feb 1987 (Annapolis, 1987); Susan Worsham, "CNO Lauds MSO's," *Surface Warfare* 7 (Jun 1982): 37; MacDonald, "New Surface MCM Ships," p. 24; John H. Hubbard, "Mine Warfare Officer Assignments: Looking Good," *Surface Warfare* 7 (Feb 1982): 19; "US and UK Coastal Minehunter Developments," *Maritime Defence* 9 (Jan 1984): 24.

52. Heine, "Sweeping Ahead," p. 9; Preston, "The Infernal Machine," p. 562; Lobb, "Mine Countermeasures," p. 35; Nepean, "Naval Mine Countermeasures," p. 116; Kenneth Keller, "MCM Avenger Class: Learning the New Technology," *Surface Warfare* 12 (May-Jun 1987): 14; USNI Seminar, "Mine Warfare."

53. *Naval Mine Warfare*, a supplement to *International Defense Review* 22 (Nov 1989), p. 14.

54. Mel R. Jones, "Closing the Mine Warfare Gap," *Proceedings* 110 (Nov 1984): 152; Jones, "Mine Warfare Preparedness," pp. 37–39; Friedman, "US Mine-Countermeasures Programs," pp. 1259–64; Polmar, "The U.S. Navy," p. 119; John D. Alden, "Tomorrow's Fleet," *Proceedings* 113 (May 1987): 180; Scott C. Truver, "Tomorrow's Fleet," *Proceedings* 115 (May 1989): 311. The MCM

is equipped with the advanced Raytheon AN/SQQ–32 minehunting sonar, the remote Honeywell Mine Neutralization System (MNS), and the AN/SSN–2 Precise Integrated Navigation System (PINS). Friedman, "US Mine-Countermeasures Programs," p. 1259.

55. Testimony of Everett Pyatt, Assistant Secretary of the Navy (Shipbuilding and Logistics), HASC Hearing, pp. 2, 6–7, 25 .

56. Thomas H. Moorer, "Mine Warfare: Entering a New Age," *Wings of Gold* 9 (Fall 1984): 8; Emshwiller Comment, p. 78. The Navy's AMCM squadrons consist of the original HM–12, now a training squadron, and two operational units, HM–14 on the East Coast and HM–16 on the West Coast.

57. Clawson, "Sweeping the Med," p. 17.

58. Action Information Organisation for MCMV, "Mine Countermeasures," *Navy International* 89 (Jul 1984): 404; Moorer, "Mine Warfare," pp. 7–8; USNI Seminar, "The Gulf of Suez Mining Crisis: Terrorism at Sea," 30 May 1985, pp. 3–22; Preston, "The Infernal Machine," p. 561; *Navy Times*, 20 Aug 1984; Michael Collins Dunn, "Fishing in Troubled Waters," *Defense and Foreign Affairs* 12 (Oct 1984): 16–18; Scott C. Truver, "Mines of August: An International Whodunit," *Proceedings* 111 (May 1985): 95–102. As many as nineteen ships may have been mined, although some vessels sustained injuries of suspicious origin.

59. USNI Seminar, "The Gulf of Suez Mining Crisis," pp. 3–22; Preston, "The Infernal Machine," p. 561; Dunn, "Fishing in Troubled Waters," pp. 16–18; Truver, "Mines of August," pp. 102–14; USNI Seminar, "Mine Warfare." The sleds used were the Mk 106, a combination of the Mk 105 magnetic sled with an Mk 104 acoustic actuator trailing.

60. USNI Seminar, "The Gulf of Suez Mining Crisis," pp. 3–22; Preston, "The Infernal Machine," p. 561; Dunn, "Fishing in Troubled Waters," pp. 16–18; Truver, "Mines of August," pp. 102–14; *Navy Times* 7 Aug 1984, p. 2; 22 Oct 1984, p. 33. As many as thirty mines were swept by all forces. USNI Seminar, "Mine Warfare."

61. Truver, "Mines of August," pp. 113–14.

62. *Congressional Quarterly* 46 (23 Apr 1988): 5–6; Thomas Q. Donaldson, "Meandering Mines," *Proceedings* 110 (Sep 1984): 137; Friedman, *World Naval Weapons Systems*, pp. 451, 478.

63. [Paul Stillwell], "SS *Bridgeton*: The First Convoy: Interview with Captain Frank C. Seitz, Jr., U.S. Merchant Marine," *Proceedings* 114 (May 1988): 52, 56; RADM W. W. Mathis to author, 21 Aug 1989; Administrative Support Unit (ASU), Bahrain, Navy Commendation Medals File; USS *Wainwright* to SECDEF et al., 160758Z APR 88, enclosure to *Samuel B. Roberts* Command History, 1988, SH. At the time of *Bridgeton*'s mining, Captain William W. Mathis, commanding *Fox*, was leading the convoy 1000 yards off *Bridgeton*'s starboard bow. Mathis to author, 21 Aug 1989. In May 1989 ASU, Bahrain, submitted recommendations for Navy Commendation Medals for the first *Hunter* and *Striker* crews, which served from August to November 1987. *Striker* crew members included BMC W. J. Walters, QMC L. J. Tebbetts, BM1 G. W. Barnhart, OS1 W. E. Chambers, SN C. E. Bush, and FN W. J. Whitmore; *Hunter* crew members were LT M. E. Magil, BMC J. W. Sharbutt, EM1 J. L. Moore, EN2 E. I. Baylock, and SN T. A. Blasier. ASU, Bahrain, Navy Commendation Medals File. Explosive Ordnance Disposal (EOD) forces included detachments from the newly established Explosive Ordnance Disposal Mobile Unit (EODMUG) in Charleston, South Carolina.

64. *Washington Post*, 25 Nov 1987, p. A20; HM–14 Lessons Learned, encl. 2, OA. *Inflict* and *Fearless* found another field of thirteen mines in the same spot four months later.

65. Vice Admiral Hogg Testimony, HASC Hearing, p. 13; HM–14 Lessons Learned, encl. 2, OA; Friedman, *World Naval Weapons Systems*, p. 475. On this recent deployment four hundred personnel supported the operations of eight aircraft. Air crews alone required twenty to twenty-five men per helicopter. HM–14 Lessons Learned, encl. 2, OA. One MSB, equipped with a jury-rigged forward-looking sonar, failed to detect mines except in very stable water. Three

MSBs with mechanical sweep gear were used in Bahrain Bell but failed to sweep mines in that environment.

66. Simmons, "Mining at Wonsan and in the Persian Gulf," p. 3; James D. Hessman, "Mine Warfare: A Sweeping Assessment," *Sea Power* 30 (Sep 1987): 7; Harold Evans, "Missing Mine Sweepers," *U.S. News and World Report*, 24 Aug 1987, p. 68; Michael Mecham, "Navy Deploys Helicopters to Counter Gulf Mine Threat," *Aviation Week and Space Technology* (3 Aug 1987): 25; Giusti, "Sweeping the Gulf," pp. 4–5. North Korea is believed to be the supplier of Iranian mines. Ibid., p. 26. Some MSBs also swept prior to the arrival of the MSOs. Giusti, "Sweeping the Gulf," p. 3; Heine, "Sweeping Ahead," pp. 7–8. *Illusive* was the only ship of the six not from the reserve fleet. Although the Atlantic Fleet ships were towed to the gulf, they returned to the United States under their own power in March 1989, while Pacific Fleet MSOs remained in the gulf until 1990. *Conquest* also suffered a collision at sea with *Barbour County* (LST–1195) on 10 September 1987 and had to return to Pearl Harbor for repairs, delaying her entry into the gulf. Pacific and Atlantic MSO crews also rotated differently with one-fourth of the Pacific Fleet crews rotating each month, whereas the Atlantic Fleet held entire crew teams to a four-month rotation. For more information on the MSO Persian Gulf experience, see individual ships' Command Histories, 1987–1990, SH.

67. USNI Seminar, "Mine Warfare"; Giusti, "Sweeping the Gulf," pp. 3–5; Heine, "Sweeping Ahead," p. 7; Friedman, *World Naval Weapons Systems*, p. 446; Walters, "ROVs," p. 301; John Marriott, "Report on the Royal Navy Equipment Exhibition," *Armada* 6 (Nov-Dec 1987): 74–75; *Washington Times*, 5 Apr 1989, p. 2. Witnesses to the failed MSO modernization program of the late 1960s no doubt were surprised by the good performance of the surviving vessels of this class in the Arabian Gulf.

68. Admiral Trost's remarks, COMINEWARCOM change of command, 8 Jul 1989; Vice Admiral Hogg Testimony, HASC Hearing, p. 18; CMEF Command History (unclassified portions), 1988, OA. Although generally too lightweight for Arabian Gulf operations, the ROVs used by the U.S. Navy MCM forces in the gulf did find some mines. Msg, CHINFO Washington DC to CJTFME, 160015Z Apr 1988, enclosure to *Samuel B. Roberts* Command History, 1988, SH; *Enhance* Command History, 1988, SH.

69. Msg, CHINFO Washington DC to CJTFME, 160015Z Apr 1988; Preston, "The Infernal Machine," p. 564; Steven Blaisdell, "HSL–44 Detachment Five Had a Blast in the Persian Gulf," *Rotor Review* (Aug 1988): 55.

70. Master Chief Bobby Scott and Boatswain's Mate Wilfred Patnaude, USS *Iowa*, taped interviews with author, 30 Mar 1988, NHC, OA; Norman Friedman, "World Naval Developments," *Proceedings* 114 (Jun 1988): 119. Captain Frank Lugo on the MEF staff during *Iowa*'s deployment also considered putting bowsweep equipment on his cruiser, *Josephus Daniels* (CG–27) for her 1988 Arabian Gulf cruise. Lugo, interview with author, 6 Nov 1988, on board *Josephus Daniels*, Mina Sulman Pier, Manama, Bahrain.

71. James R. Giusti, "Sweeping the Gulf," *Surface Warfare* 13 (Mar-Apr 1988): 3; Michael Chichester, "Allied Navies & the Gulf War: Strategic Implications," *Navy International* 93 (Jun 1988): 320. The contact mines captured aboard the *Iran Ajr* may be Russian-supplied from North Korean stock. Preston, "The Infernal Machine," p. 559; A. H. Cordesman, "Western Seapower Enters the Gulf (Part 2)," *Naval Forces* 9 (1988): 40.

72. Msgs, USS *Wainwright* to SECDEF et al., 160758Z Apr 1988, USS *Jack Williams* to NMCC Washington DC et al., 172000Z Apr 1988, USCINCCENT to CJTFME, 141920Z Apr 1988, USS *Wainwright* to CJTFME, 161224Z Apr 1988, and CHINFO Washington DC to CJTFME, 160015Z Apr 1988, enclosures to *Samuel B. Roberts* Command History, 1988; Chuck Mussi, "'To See the Dawn:' The Night-long Battle to Save USS *Roberts*," *All Hands* 857 (Aug 1988): 4–10. The intended victim of the mining attack was probably *Bridgeton*, due to proceed on Earnest Will mission EW 88026, scheduled for 14 April. Msgs, USCINCCENT to CJTFME, 141920Z Apr 1988,

and USS *Wainwright* to CJTFME 1612242 Apr 1988, enclosures to *Samuel B. Roberts* Command History, 1988. For a description of the action on 18 April 1988, see "Awards for Persian Gulf Actions," *Surface Warfare* 14 (May-Jun 1989): 17.

73. "Desert Shield, Getting There," *Proceedings* 116 (Oct 1990): 102–3.

74. *Washington Post*, 19 Feb 1991, p. A7; *Navy Times*, 4 Mar 1991, pp. 10–11, 18.

75. *Navy Times*, 4 Mar 1991, pp. 10–12; ibid., 11 Mar 1991, p. 12. *Guardian* (MCM–5) relieved *Avenger* in June 1991 and found several mines in the last months of gulf MCM efforts.

76. R. J. L. Dicker, "Mine Warfare Now and in the 1990s," *International Defense Review* 19 (1986): 294; Vice Admiral Hogg Testimony, HASC Hearing, pp. 11–13; David C. Resing, "Mine Countermeasures in Coastal Harbors: A Force Planner's Dilemma," *Naval War College Review* 40 (Spring 1987): 60; HM–14 Lessons Learned, encl. 2, OA. Figures for new construction change daily. At last report, the CNO had requested 22 MCMs, 25 MHCs, and 44 helicopters. Budget figures comparing MCM and USN budgets are for the fiscal year 1990, 1991, and 1992, the highest MCM budget years since the Korean War.

Conclusion

1. Wettern, "Mine Countermeasures," p. 21.

2. Edward J. Rogers, "Mines Wait But We Can't," *Proceedings* 108 (Aug 1982): 52.

3. USNI Seminar, "The Gulf of Suez Mining Crisis: Terrorism at Sea," p. 18.

4. Remarks by Admiral Trost at COMINEWARCOM change of command, 8 Jul 1989.

5. Daniel, *Anti-Submarine Warfare*, pp. 135, 169; Truver, "Weapons That Wait," p. 33; Friedman, "US Mine-Countermeasures," p. 1268; White, "Move and Countermove," p. 35.

6. Ennis, "Deep, Cheap & Deadly," p. 26; Norman Polmar, *Guide to the Soviet Navy* (Annapolis, 1986), pp. 248–561; Chatterton, *The Minesweeping/Fishing Vessel*, p. 121; Mel R. Jones, "Mine Warfare Preparedness Begins Recovery After Years of Neglect," *Defense Systems Review and Military Communication* 2 (Jul-Aug 1984): 37; Bray, "Mine Warfare in the Russian and Soviet Navies," pp. 2, 39–40, 177–79; Wile, "Their Mine Warfare Capability," p. 149.

7. For unclassified technical information on the mines and MCM devices of all nations, see Friedman, *World Naval Weapons Systems*, pp. 444–79.

8. James H. Patton, Jr., "ASW: Winning the Race," *Proceedings* 114 (Jun 1988): 66.

9. Mine warfare type commanders have repeatedly been dual-hatted as operational commanders since World War II. For example, see Rear Admiral Sharp (chapter 2) and Rear Admiral McCauley (twice in chapter 4). The 1968 MINEPAC Command History also lists COMINEPAC serving double duty as CTF 59 during that year. Undoubtedly, there are other examples.

10. Hoffman, "Offensive Mine Warfare," p. 155; USNI Seminar, "Mine Warfare."

11. Daniel G. Powell Comment, *Proceedings* 106 (Jun 1980): 79.

Select Bibliography

Document Collections

Hagley Museum and Library. Wilmington, DE.

   Alexander McKinley Diary

Library of Congress. Manuscript Division. Washington, DC.

   David G. Farragut Papers
   John Crittenden Watson Papers

National Archives and Records Administration. Washington, DC.

   Record Group 24:
        Log of USS *Cowslip*
        Log of USS *Hartford*
        Log of USS *Monongahela*
        Log of USS *Sebago*

   Record Groups 38, 80, and 313:
        Records Relating to United States Navy Fleet Problems I to XXII,
        1923–1941

   Record Group 80:
        Records of the General Board

Naval Historical Center. Washington, DC.

   Naval Aviation History:
        Command Histories

   Operational Archives:
        World War II Action Reports
        World War II Command File
        Post-1 Jan 1946 Command File
        Post-1 Jan 1974 Command File
        U.S. Naval Forces, Vietnam, Monthly Historical Summaries

   Ships Histories Branch:
        Command Histories

Naval Historical Foundation. Washington, DC.

   David G. Farragut Papers

Bibliography

Documents

Bray, Jeffrey K. "Mine Warfare in the Russian and Soviet Navies." M.A. Thesis, Old Dominion University, 1989.

Catlin, George L. "Paravanes: A History of the Activities of the Bureau of Construction and Repair in Connection With Paravanes in the War With Germany." Unpublished manuscript, Navy Department Library, 1919.

Chief of Naval Operations. "Mine Warfare in the Naval Establishment." Washington: n.d.

______. Mine Warfare Section. *Mine Warfare Notes*. World War II Command File, Operational Archives, Naval Historical Center.

______. Mine Warfare Section. "U.S. Navy Mine and Bomb Disposal in World War II." Unpublished manuscript, World War II Command File, Operational Archives, Naval Historical Center.

______. Naval Inshore Warfare Command, Atlantic. "Final Report, Suez Canal Clearance Operation, Task Force 65." May 1975. Post-1 Jan 1974 Command File, Operational Archives, Naval Historical Center.

Feldman, D. W. "Historical Review of Mine Countermeasures Techniques and Devices." Technical Note No. 449. Mine Countermeasures Department, Naval Coastal Systems Center, Panama City, FL.

Fleet and Mine Warfare Training Center, Charleston, SC. "Fundamentals of Mine Warfare Planning." (Confidential) Course No. A–2A–0010. Charleston: 10 June 1975.

______. "Surface Mine Warfare Familiarization." (Confidential) Courses No. A–2E–0017 and A–2E–0033. Charleston: 1 October 1973.

Gruendl, Paul L. "U.S. Navy Airborne Mine Countermeasures: A Coming of Age." Professional Study No. 5617. Maxwell Air Force Base, AL: Air War College, 1975.

Hart, Thomas C. *Narrative of Events, Asiatic Fleet leading up to War and from 8 December 1941 to 15 February 1942*. World War II Action Reports, Operational Archives, Naval Historical Center.

Hartmann, Gregory K. "Mine Warfare History and Technology." Naval Surface Weapons Center Technical Report 75–88. White Oak, MD: Naval Surface Weapons Center, 1 July 1975.

______. *Wave Making by an Underwater Explosion, NSWC / WOL MP 76–15*. White Oak, MD: Naval Surface Weapons Center, 1976.

Low, Francis S. "Study of Undersea Warfare." 22 April 1950. Post-1 Jan 1946 Command File, Operational Archives, Naval Historical Center.

McEathron, Ellsworth Dudley. "Minecraft in the Van: A History of United States Minesweeping in World War II." 3 vols. Typescript, World War II Command File, Operational Archives, Naval Historical Center.

Massachusetts Institute of Technology. "Project Hartwell: A Report on Security of Overseas Transport." 2 vols. 21 September 1950. Post-1 Jan 1946 Command File, Operational Archives, Naval Historical Center.

Mine Advisory Committee. National Academy of Sciences, National Research Council. Project Nimrod. "The Present and Future Role of the Mine in Naval Warfare." Chapter 2. Washington: National Academy of Sciences, 15 January 1970. Post-1 Jan 1946 Command File, Operational Archives, Naval Historical Center.

Moorer, Thomas H. Speeches and Statements by Admiral Thomas H. Moorer, Chief of Naval Operations, United States Navy. Washington, D.C., 1 August 1967–1 July 1970. Special Collections, Navy Department Library.

Naval Sea Systems Command. "Operation 'End Sweep': A History of Mine Sweeping Operations in North Vietnam, 8 May 1972–28 July 1973." (Sanitized) Operational Archives, Naval Historical Center.

Pacific Fleet, Commander in Chief. *Korean War. U.S. Pacific Fleet Operations. Mine Warfare.* Interim Evaluation Reports. Operational Archives, Naval Historical Center.

Pittsburgh, University of. Historical Staff at ONR. "The History of the United States Naval Research and Development in World War II." Typescript, Navy Department Library, 1949.

U.S. Bureau of Construction and Repair. Navy Department. *Otter Handbook for Merchant Ships.* Washington: GPO, 1918.

U.S. Bureau of Naval Personnel. Navy Department. *The Evolution of Naval Weapons.* Washington: GPO, 1947.

U.S. Bureau of Naval Weapons. Navy Department. *Demolition Materials.* Washington: GPO, 1962.

U.S. Bureau of Navigation. Navy Department. *Mine-Sweeping Manual.* Washington: GPO, 1917.

U.S. Bureau of Ordnance. Navy Department. *U.S. Navy Bureau of Ordnance in World War II.* Washington: GPO, 1953.

Bibliography

U.S. Congress. House. Committee on Armed Services. *Mine Warfare: Hearing Before the Seapower and Strategic and Critical Materials Subcommittee.* 100th Cong., 1st Sess., 17 September 1987, H.A.S.C. No. 100–50.

U.S. Department of Defense. *Semiannual Report of the Secretary of Defense and the Semiannual Reports of the Secretary of the Army, Secretary of the Navy, Secretary of the Air Force.* Washington: GPO, various years.

U.S. Naval Academy. Department of Ordnance and Gunnery. *Naval Ordnance and Gunnery.* Vol. 1. Reprint. Washington: GPO, 1957.

U.S. Naval Education and Training Command. *Mineman 1 & C.* Washington: U.S. Naval Education and Training Command, 1983.

U.S. Naval Institute Professional Seminar Series. *The Gulf of Suez Mining Crisis: Terrorism at Sea.* Annapolis: U.S. Naval Institute, 30 May 1985.

______. *Mine Warfare: Which Platform?* Annapolis: U.S. Naval Institute, 26 February 1987.

U.S. Naval Oceanographic Office. *DAPAC: Danger Areas in the Pacific.* Washington: GPO, 1967.

U.S. Naval Torpedo Station. *Mines and Countermines, U.S.N.* [Newport, RI: Naval Torpedo Station], 1902.

U.S. Navy Department. *Official Records of the Union and Confederate Navies in the War of the Rebellion.* Washington: GPO, 1894–1913.

______. North Sea Minesweeping Detachment. *Sweeping the North Sea Mine Barrage.* New York: North Sea Minesweeping Detachment, 1919.

______. Office of Naval Records and Library. *The Northern Barrage and Other Mining Activities.* Washington: GPO, 1920.

______. *The Northern Barrage: Taking Up the Mines.* Washington: GPO, 1920.

______. Secretary of the Navy Annual Report. *Semiannual Report of the Secretary of Defense.* Washington: various years.

______. United States Naval Administration in World War II. Bureau of Ordnance, Miscellaneous Activities. Vol. 1, "A History of the U.S. Naval Mine Warfare Test Station, Solomons, Maryland." Typescript, Navy Department Library.

______. United States Naval Administration in World War II. Bureau of Ordnance. "Underwater Ordnance." Typescript, Navy Department Library.

______. United States Naval Administration in World War II. Commander Minecraft, U.S. Pacific Fleet. "Administrative History of Minecraft, Pacific Fleet." Typescript, Navy Department Library.

______. United States Naval Administration in World War II. "History of the Naval Ordnance Laboratory, 1918–1945." Typescript, Navy Department Library.

U.S. Strategic Bombing Survey. *The Offensive Mine Laying Campaign Against Japan.* 1946. Reprint. Washington: Naval Material Command, 1969.

U.S. War Department. *The War of the Rebellion: A Compilation of the Official Records of the Union and Confederate Armies.* 128 vols. Washington: GPO, 1880–1901.

Watson, John C. "Farragut and Mobile Bay, Personal Reminiscences." *War Paper No. 98.* Paper delivered at the meeting of the Military Order of the Loyal Legion of the United States, Commandery of the District of Columbia, 16 December 1916.

Winters, Robert. "History and Analysis: Mine Warfare Operations in Southeast Asia, 1961–1973." Post-1 Jan 1946 Command File, Operational Archives, NHC.

Books

[All Hands]. *The Northern Barrage: Mine Force, United States Atlantic Fleet.* Annapolis: U.S. Naval Institute, 1919.

Abbot, H. L. *Beginnings of Modern Submarine Warfare.* Willets Point, NY: Battalion Press, 1881.

Anderson, Jane, and Gordon Bruce. *Flying, Submarining and Mine Sweeping.* London: Sir J. Causton and Sons, 1916.

Auer, James E. *The Postwar Rearmament of Japanese Maritime Forces, 1945–71.* New York: Praeger Publishers, 1973.

Barnes, J. S. *Submarine Warfare-Offensive and Defensive.* New York: Van Nostrand, 1869.

Bearss, Edwin C. *Hardluck Ironclad: The Sinking and Salvage of the Cairo.* Baton Rouge: Louisiana State University Press, 1966.

Belknap, Reginald Rowan. *The Yankee Mining Squadron, or Laying the North Sea Mine Barrage.* Annapolis: U.S. Naval Institute, 1920.

Bradford, R. B. *History of Torpedo Warfare.* Newport, RI: U.S. Torpedo Station, 1882.

Bibliography

Breemer, Jan S. *U.S. Naval Developments*. Annapolis: Nautical and Aviation Publishing Company of America, 1983.

Brodie, Bernard. *Sea Power in the Machine Age*. Princeton: Princeton University Press, 1941.

Bucknill, J. T. *Submarine Mines and Torpedoes as Applied to Harbor Defense*. New York: Wiley, 1889.

Bush, Vannevar. *Modern Arms and Free Men*. New York: Simon and Schuster, 1949.

Cagle, Malcolm W., and Frank A. Manson. *The Sea War in Korea*. Annapolis: U.S. Naval Institute, 1957.

Canfield, Eugene B. *Notes on Naval Ordnance of the American Civil War, 1861–1865*. Washington: American Ordnance Association, 1960.

Christman, Albert B. *Naval Innovators 1776–1900*. Dalgren, VA: Naval Surface Weapons Center, 1989.

Clark, Thomas. *D. Bushnell, Inventor of the Torpedo, etc*. Philadelphia, 1814.

Conway's All the World's Fighting Ships, 1922–1946. New York: Mayflower Books, 1980.

Conway's All the World's Fighting Ships, 1947–1982, Part 1, *The Western Powers*. Annapolis: Naval Institute Press, 1983.

Cowie, J. S. *Mines, Minelayers and Minelaying*. New York: Oxford University Press, 1949.

Craighill, William. *Foreign Systems of Torpedoes as Compared With Our Own*. Willets Point, NY: Battalion Press, 1888.

Croziat, Victor. *The Brown Water Navy: The River and Coastal War in Indo-China and Vietnam, 1948–1972*. Dorset, UK: Blandford Press, 1984.

Curtis, John Shelton. *Russia's Crimean War*. Durham: Duke University Press, 1979.

Cutler, Thomas J. *Brown Water, Black Berets: Coastal and Riverine Warfare in Vietnam*. Annapolis: Naval Institute Press, 1988.

Daniel, Donald C. *Anti-Submarine Warfare and Superpower Strategic Stability*. Chicago: University of Illinois Press, 1986.

Daniels, Josephus. *Our Navy at War*. Washington: Pictorial Bureau, 1922.

Denlinger, Sutherland, and Charles B. Gary. *War in the Pacific: A Study of Navies, Peoples, and Other Battle Problems*. New York: R. M. McBride & Co., 1936.

Dewey, George. *Autobiography of George Dewey, Admiral of the Navy*. New York: Charles Scribner's Sons, 1913.

Dewitz, H. von. *War's New Weapons: An Expert Analysis in Plain Language of the Weapons and Methods Used in the Present Great War*. New York: Dodd, Mead & Co., 1915.

Dickinson, H. W. *Robert Fulton, Engineer and Artist: His Life and Works*. London: J. Lane, 1913.

Dillard, Walter Scott. *Sixty Days to Peace: Implementing the Paris Peace Accords*. Washington: National Defense University, 1982.

Dommett, W. E. *Submarine Vessels, Including Mines, Torpedoes, Guns, Steering, Propelling*. London: Whittaker & Co., 1915.

Domville-Fife, Charles William. *Submarines, Mines and Torpedoes in the War*. London and New York: Hodder & Stoughton, 1914.

Dorling, Henry Taprell. *Swept Channels: Being an Account of the Work of the Minesweepers in the Great War*. London: Hodder & Stoughton, 1935.

Duncan, Robert C. *America's Use of Sea Mines*. White Oak, MD: U.S. Naval Ordnance Laboratory, 1962.

Edwards, Kenneth. *The Navy of To-day*. London: Blackie & Son, 1939.

Elliott, Peter. *Allied Minesweeping in World War II*. Annapolis: Naval Institute Press, 1979.

Field, James A., Jr. *History of U.S. Naval Operations: Korea*. Washington: Naval History Division, 1962.

Friedman, Norman. *The Postwar Naval Revolution*. Annapolis: Naval Institute Press, 1986.

_____. *The US Maritime Strategy*. New York: Jane's Publishing, 1988.

_____. *World Naval Weapons Systems*. Annapolis: U.S. Naval Institute Press, 1989.

Bibliography

Fulton, Robert. *Torpedo War and Submarine Explosions*. New York: W. Eliot, 1810.

Fulton, William B. *Riverine Operations, 1966–1969*. Washington: Department of the Army, 1973.

Furer, Julius Augustus. *Administration of the Navy Department in World War II*. Washington: GPO, 1959.

Gannon, Michael. *Operation Drumbeat*. New York: Harper & Row, 1990.

Gerken, Louis C. *Mine Warfare Technology*. Chula Vista, CA: American Scientific Corp., 1989.

Goode, W. A. M. *With Sampson Through the War*. New York: Doubleday & McClure Co., 1899.

Grant, Robert M. *U-boats Destroyed: The Effect of Anti-Submarine Warfare, 1914–1918*. London: Putnam, 1964.

Great Britain. Admiralty. *His Majesty's Minesweepers*. London: HMSO, 1943.

Gregory, Barry. *Vietnam Coastal and Riverine Forces Handbook*. Irthlingborough, UK: Patrick Stephens, 1988.

Griffiths, Maurice. *The Hidden Menace*. Greenwich: Conway Maritime Press, 1981.

Grossnick, Roy A., ed. *Kite Balloons to Airships. . . The Navy's Lighter-Than-Air Experience*. Washington: Naval Historical Center, 1986.

Grosvenor, Joan. *Open the Ports: The Story of Human Minesweepers*. London: W. Kimber, 1956.

Hall, Cyril. *Modern Weapons of War, by Land, Sea, and Air*. London: Blackie & Son, 1916.

Hartmann, Gregory Kemenyi. *Weapons That Wait: Mine Warfare in the U.S. Navy*. Annapolis: Naval Institute Press, 1979.

Hastings, Max. *The Korean War*. New York: Simon & Schuster, 1987.

Hooper, Edwin Bickford. *Mobility, Support, Endurance: A Story of Naval Operational Logistics in the Vietnam War, 1965–1968*. Washington: Naval History Division, 1972.

[Hooper, Edwin B.]. *The Navy Department: Evolution and Fragmentation.* Washington: Naval Historical Foundation, 1978.

Hooper, Edwin Bickford, Dean C. Allard, and Oscar P. Fitzgerald. *The United States Navy and the Vietnam Conflict. Vol. 1, The Setting of the Stage to 1959.* Washington: Naval History Division, 1976.

Hubbe, M. *Submarine Warfare: Fixed Mines and Torpedoes. The Lay Moveable Torpedo.* Buffalo: n.d.

Hutcheon, Wallace. *Robert Fulton, Pioneer of Undersea Warfare.* Annapolis: Naval Institute Press, 1981.

International Symposium on Mine Warfare Vessels and Systems. *Mine Warfare Vessels and Systems.* London: The Royal Institution of Naval Architects, 1984

Ito, Masanori, with Roger Pineau. *The End of the Imperial Japanese Navy.* New York: Macfadden-Bartell, 1965.

Jellicoe, John R. *The Crisis of the Naval War.* New York: George H. Deran Co., 1921.

Johnson, Ellis A., and David A. Katcher. *Mines Against Japan.* White Oak, MD: Naval Ordnance Laboratory, 1973.

Jones, Virgil Carrington. *The Civil War at Sea, July 1863-November 1865: The Final Effort.* New York: Holt, Rinehart, Winston, 1962.

Karnecke, Joseph Sidney. *Navy Diver.* New York: Putnam, 1962.

Lenton, H. T. *American Gunboats and Minesweepers.* New York: Arco Publishing, 1974.

Lewis, Charles Lee. *David Glasgow Farragut: Our First Admiral.* Annapolis: U.S. Naval Institute, 1943.

Lott, Arnold S. *Most Dangerous Sea: A History of Mine Warfare and an Account of U.S. Navy Mine Warfare Operations in World War II and Korea.* Annapolis: U.S. Naval Institute, 1959.

Love, Robert William, Jr., ed. *The Chiefs of Naval Operations.* Annapolis: Naval Institute Press, 1980.

Low, Archibald Montgomery. *Mine and Countermine.* New York: Sheridan House, 1940.

Bibliography

Lund, Paul. *Out Sweeps! The Story of the Minesweepers in World War II.* London and New York: Foulsham, 1978.

Lundeberg, P. K. *Samuel Colt's Submarine Battery: the Secret and the Enigma.* Smithsonian Studies in History and Technology, No. 29. Washington: Smithsonian Institution Press, 1979.

Mackay, Ruddock F. *Fisher of Kilverstone.* Oxford: Clarendon Press, 1973.

Marolda, Edward J., and Oscar P. Fitzgerald. *The United States Navy and the Vietnam Conflict.* Vol. 2. *From Military Assistance to Combat, 1959–1965.*Washington: Naval Historical Center, 1986.

Marolda, Edward J., and G. Wesley Pryce III. *A Short History of the United States Navy and the Southeast Asian Conflict, 1950–1975.* Washington: Naval Historical Center, 1984.

Morison, Samuel Eliot. *Aleutians, Gilberts and Marshalls, June 1942–April 1944.* Boston: Little, Brown & Co., 1961.

______. *The Invasion of France and Germany, 1944–1945.* Boston: Little, Brown & Co., 1957.

______. *Leyte, June 1944-January 1945.* Boston: Little, Brown & Co., 1961.

______. *The Liberation of the Philippines: Luzon, Mindanao, the Visayas, 1944–1945.* Boston: Little, Brown & Co., 1969.

______. *New Guinea and the Marianas, March 1944-August 1944.* Boston: Little, Brown & Co., 1961.

______. *Operations in North African Waters, October 1942-June 1943.* Boston: Little, Brown & Co., 1947.

______. *Sicily–Salerno–Anzio, January 1943-June 1944.* Boston: Little, Brown & Co., 1962.

______. *Victory in the Pacific, 1945.* Boston: Little, Brown & Co., 1961.

Naval History Division. *Dictionary of American Naval Fighting Ships.* 8 vols. Washington: Naval History Division, 1970–.

Nevinson, Henry W. *The Dardanelles Campaign.* New York: Henry Holt & Co., 1919.

Palmer, Michael A. *Origins of the Maritime Strategy: American Naval Strategy in the First Postwar Decade.* Washington: Naval Historical Center, 1988.

Parker, Foxhall A. *The Battle of Mobile Bay, and the Capture of Forts Powell, Gaines and Morgan, by the Combined Sea and Land Forces of the United States, Under the Command of Rear-Admiral David Glasgow Farragut, and Major-General Gordon Granger, August, 1864.* Boston: A. Williams & Co., 1878.

Parsons, William B. *Robert Fulton and the Submarine.* New York: Columbia University Press, 1922.

Patterson, Andrew, ed. *Historical Bibliography of Sea Mine Warfare.* Washington: National Academy of Sciences, 1977.

Perry, Milton F. *Infernal Machines: The Story of Confederate Submarine and Mine Warfare.* Baton Rouge: Louisiana State University Press, 1965.

Polmar, Norman. *Guide to the Soviet Navy.* Annapolis: Naval Institute Press, 1986.

_____. *Ships and Aircraft of the U.S. Fleet.* Annapolis, MD: U.S. Naval Institute, 1985.

_____. *Soviet Naval Power: Challenge for the 1970s.* New York: National Strategy Information Center, 1974.

Potter, E. B. *Admiral Arleigh Burke.* New York: Random House, 1990.

Potter, E. B., and Chester W. Nimitz, eds. *Sea Power: A Naval History.* Englewood Cliffs, NJ: Prentice-Hall, 1960.

Preston, Antony. *Navies of World War 3.* New York: Military Press, 1984.

Ranft, Bryan, and Geoffrey Till. *The Sea in Soviet Strategy.* Annapolis: Naval Institute Press, 1983.

Read, J. *Explosives.* New York: Penguin Books, 1943.

Reigart, J. F. *Life of Robert Fulton.* Philadelphia: C. G. Henderson & Co., 1856.

Rickover, H. G. *How the Battleship Maine Was Destroyed.* Washington: Naval History Division, 1976.

Roland, Alex. *Underwater Warfare in the Age of Sail.* Bloomington: Indiana University Press, 1978.

Rowland, Buford, and William B. Boyd. *U.S. Navy Bureau of Ordnance in World War II.* Washington: Department of the Navy, n.d.

Bibliography

Roye, John S., and Samuel L. Morison. *Ships and Aircraft of the U.S. Fleet.*
Annapolis: Naval Institute Press, 1972.

Saga Shannon: The Story of the USS SHANNON DM25 in Action, 1944–1946. New
York: J. Curtis Blue, 1947.

Scheliha, Viktor Ernst Karl Rudolf von. *A Treatise on Coast-Defence: Based on the
Experience Gained by Officers of the Corps of Engineers of the Army of the
Confederate States, and Compiled From Official Reports of the Officers of the
Navy of the United States, Made During the Late North American War from
1861 to 1865.* London: E. & F. N. Spon, 1868.

Schubert, Paul. *Sea Power in Conflict.* New York: Coward-McCann, 1942.

Sims, William Snowden. *The Victory at Sea.* Garden City, NY: Doubleday, Page &
Co., 1920.

Sleeman, C. W. *Torpedoes and Torpedo Warfare: Containing a Complete and Concise
Account of the Rise and Progress of Submarine Warfare; Also a Detailed
Description of all Matters Appertaining Thereto, Including the Latest
Improvements.* Portsmouth, UK: Griffin, 1880.

Spector, Ronald. *Admiral of the New Empire: The Life and Career of George Dewey.*
Baton Rouge: Louisiana State University Press, 1974.

Stickney, Joseph L. *Admiral Dewey at Manila.* Philadelphia, PA: J. H. Moore Co.,
1899.

Sueter, Murray F. *The Evolution of the Submarine Boat, Mine and Torpedo: From
the Sixteenth Century to the Present Time.* Portsmouth, UK: J. Griffin & Co.,
1908.

Trask, David F. *The War With Spain in 1898.* New York: Macmillan Publishing Co.,
1981.

Turner, John Frayn. *Service Most Silent: The Navy's Fight Against Enemy Mines.*
London: Harrap, [1955].

Wallin, Jeffrey D. *By Ships Alone: Churchill and the Dardanelles.* Durham, NC:
North Carolina Academic Press, 1981.

Weigley, Russell F. *The American Way of War: A History of United States Military
Strategy and Policy.* Bloomington: Indiana University Press, 1973.

West, Richard S. *Admirals of American Empire.* New York: Bobbs-Merrill Co., 1948.

White, Peter B. *Defending a Logistics System Under Mining Attack*. Arlington, VA: Center for Naval Analyses, 1976.

Wilson, H. W. *The Downfall of Spain: Naval History of the Spanish-American War*. Boston: Little, Brown & Co., 1900.

Wineland, W. C. *Delay as a Measure of Minefield Effectiveness*. Naval Ordnance Laboratory Technical Report 69–206. November 1969.

Zumwalt, Elmo R., Jr. *On Watch*. New York: Quadrangle/The New York Times Book Co., 1976.

Articles

Adshead, Robin. "Minehunting." *Armed Forces* 7 (April 1988): 165–69.

"Aerial Minesweeping." *Defence Update International* (October 1986): 42–48.

Alden, John D. "The Indestructible XMAP." *Naval History* 2 (Winter 1988): 44–47.

______. "Tomorrow's Fleet." *Naval Review 1986*. U.S. Naval Institute *Proceedings* 113 (May 1987): 177–86.

Alger, P. R. "Employment of Submarine Mines in Future Wars." U.S. Naval Institute *Proceedings* 127 (1908): 1039–42.

Armstrong, Harry C. "The Removal of the North Sea Mine Barrage." *Warship International* 25 (June 1988): 134–69.

Ashton, George. "Minesweeping Made Easy." U.S. Naval Institute *Proceedings* 87 (July 1961): 66–71.

Auld, Sonny. "First LANT COOP." *Surface Warfare* 10 (November-December 1985): 10–11.

Baker, H. George. "The MSB Story: Little Ships Sweep the Sea to Keep it Free." *All Hands* 481 (February 1957): 2–5.

______. "The Mine Force: Wooden Ships and Iron Men." *All Hands* 481 (February 1957): 6–9.

Baldwin, Hanson W. " 'Spitkits' in Tropic Seas." *Shipmate* (1966). Reprinted in *Riverine Warfare: Vietnam*, pp. 56–61. Washington: Naval History Division, 1968.

Baugh, Barney. "History of Navy Mine Warfare: Kegs, Cabbages and Acoustics." *All

Bibliography

Hands 481 (February 1957): 10–15.

Beauregard, Pierre Gustave Toutant de. "Torpedo Service in the Harbor and Water Defences of Charleston." *Southern Historical Society Papers* 5 (April 1878): 145–61.

Beaver, Paul. "Aerial Minesweeping: US Navy Developments." *Navy International* 89 (May 1984): 305–9.

Bell, R., and R. Able. "Mine Warfare CO-OPeration." U.S. Naval Institute *Proceedings* 110 (October 1984): 147–49.

Blaisdell, Steven. "HSL-44 Detachment Five Had a Blast in the Persian Gulf." *Rotor Review* (August 1988): 55.

"Blockading the Blockader." *Scientific American* 116 (19 May 1917): 484.

Booda, Larry L. "Mine Warfare Moves Ahead, Plays ASW Role." *Sea Technology* 25 (November 1984): 10–17.

Booker, R. W. "Survey Ship Goes Mine Hunting: Harkness and 'Intense Look.' " *All Hands* 815 (February 1985): 32–34.

Boone, Terry, Edward Coleman, Kent Jacobson, and Joe King. "Offboard Countermeasures: Today and Tomorrow." *Defense Systems Review* 2 (July-August 1984): 40–44.

Bovey, Wilfrid. "Damn the Torpedoes. . . ?" U.S. Naval Institute *Proceedings* 65 (October 1939): 1443–51.

Bowers, John V. "Mine Warfare Channel Markers". U.S. Naval Institute *Proceedings* 89 (September 1963): 132–34.

Bowler, R. T. E. "Mine Warfare." U.S. Naval Institute *Proceedings* 92 (March 1966): 4–6.

Boyd, J. Huntly. "Nimrod Spar: Clearing the Suez Canal." U.S. Naval Institute *Proceedings* 102 (February 1976): 18–26.

Brainard, Alfred P. "Russian Mines on the Danube." U.S. Naval Institute *Proceedings* 91 (July 1965): 51–56.

Bray, Jeffrey K. "Mine Awareness." U.S. Naval Institute *Proceedings* 113 (April 1987): 41–43.

Breemer, Jan. "Mine Warfare: the Historical Perspective." *Naval Forces* 9 (1988): 36–40, 42–43.

Brookfield, S. J. "Mines and Counter-Measures." *Discovery* (January 1946): 23–29.

Bush, James C. "Development of Submarine Mines and Torpedoes." *Military Service Institute of the U.S. Journal* 11 (March, May 1890): 179–97, 377–95.

Cagle, Malcolm W., and Frank A. Manson. "Wonsan: The Battle of the Mines." U.S. Naval Institute *Proceedings* 83 (June 1957): 598–611.

Catlin, George L. "Paravanes." U.S. Naval Institute *Proceedings* 45 (July 1919): 1135–57.

Chaplin, J. B. "The Application of Air Cushion Technology to Mine Countermeasures in the United States of America." *International Symposium on Mine Warfare Vessels and Systems, London, 12-15 June 1984*. Vol. 2. London: Royal Institute of Naval Architects, 1984: 1–9.

Chatterton, Howard A. "The Minesweeping/Fishing Vessel." U.S. Naval Institute *Proceedings* 96 (June 1970): 121–25.

Chichester, Michael. "Allied Navies & the Gulf War: Strategic Implications." *Navy International* 93 (June 1988): 318–21.

Chomeau, John. "Soviet Mine Warfare." *Naval War College Review* 24 (December 1971): 94–96.

Christensen, Cyrus R. "A Minesweeping Shrimp Boat? A What?" U.S. Naval Institute *Proceedings* 107 (July 1981): 109–11.

Christian, R. C. "Minesweeping Arks." *The Fighting Forces* (June 1947): 116.

Clawson, Stephen H., and Douglas P. Tesner. "Sweeping the Med." *Surface Warfare* 7 (February 1982): 14–18.

Cockfield, D. W. "Breakout: A Key Role for the Naval Reserve." U.S. Naval Institute *Proceedings* 111 (October 1985): 52–58.

Coletta, Paolo E. "Naval Mine Warfare." U.S. Naval Institute *Proceedings* 85 (November 1959): 82–126.

______. "Naval Mine Warfare." *Navy* 2 (November 1959): 16–24.

Combs, Amy E. C. "Surface Mine Warfare." *Surface Warfare* 12 (January-February 1987): 19.

Connors, Tracy D. "The 'First Team.' " *All Hands* 858 (September 1988): 30–31.

Bibliography

"Countering the Magnetic Mine." *The Engineer* (15 March 1940). Reprinted in U.S. Naval Institute *Proceedings* 66 (May 1940): 756.

Cowie, John S. "British Mines and the Channel Dash." U.S. Naval Institute *Proceedings* 84 (April 1958): 39–47.

______. "The Mine as a Tool of Limited War." U.S. Naval Institute *Proceedings* 93 (October 1967): 106.

______. "The Role of the U.S. Navy in Mine Warfare." U.S. Naval Institute *Proceedings* 91 (May 1965): 52–63.

Cracknell, William H., Jr. "The Role of the U.S. Navy in Inshore Waters." *Naval War College Review* 21 (November 1968): 65–87.

Crane, R. H. "Mine Warfare History and Development." *Journal of the Australian Naval Institute* 10 (November 1984): 31–38.

"Current Trends in Mine Warfare." *Naval Forces: A Supplement* 6 (1984): 38–39.

Curtis, Ian. "Mines at Sea: Recognition at Last." *Defense and Foreign Affairs* 16 (July 1988): 21–24.

Daly, Thomas M. "Weekends Are Work." U.S. Naval Institute *Proceedings* 106 (August 1980): 61–65.

Davis, Noel. "The Removal of the North Sea Mine Barrage." *National Geographic Magazine* 37 (February 1920): 103–33.

Dawson, Christopher, and Mark Hewish. "Mine Warfare: New Ship Designs." *International Defense Review* 21 (August 1988): 977–86.

Deming, Angus. "A Case of the Mystery Mines." *Newsweek*, 13 August 1987, p. 46.

DeStefano, Robert. "The 27 Ship Navy: The Ultimate Mine Warfare Force?" U.S. Naval Institute *Proceedings* 114 (February 1988): 36–39.

Dicker, R. J. L. "Mine Warfare Now and in the 1990s." *International Defense Review* 19 (1986): 293–94.

Donaldson, Thomas Q. "Meandering Mines." U.S. Naval Institute *Proceedings* 110 (September 1984): 137.

Dunn, Michael Collins. "Fishing in Troubled Waters." *Defense and Foreign Affairs* 12 (October, 1984): 16–19.

Edwards, Harry William. "A Naval Lesson of the Korean Conflict." U.S. Naval Institute *Proceedings* 80 (December 1954): 1336–40.

Elsey, G. H. "Anti-mine Hovercraft: Has Their Time Come?" In *Jane's Naval Review*, edited by Captain John Moore, pp. 107–13. London: Jane's Publishing, 1986.

Ennis, John. "Deep, Cheap & Deadly." *Sea Power* 17 (December 1974): 26–30.

Evans, Harold. "Missing Mine Sweepers." *U.S. News and World Report*, 24 August 1987, p. 68.

Eveleigh, M. H. "Minesweeping." *Journal of the Royal United Service Institution* 88 (February 1943): 35–41.

Ferguson, J. N. "The Submarine Mine." U.S. Naval Institute *Proceedings* 40 (November-December 1914): 1697–1706.

Ferrand, M. C. "Torpedo and Mine Effects in the Russo-Japanese War," translated by Philip R. Alger. U.S. Naval Institute *Proceedings* 33 (December 1907): 1479–86.

"Floating Mines With Periscopes." U.S. Naval Institute *Proceedings* 161 (January-February 1916): 604.

Foisie, Jack. "Copters Sweep for Mines off Egypt." *Los Angeles Times*, 27 September 1975, p. 10.

Friedman, Norman. "Postwar British Mine Countermeasures-And National Strategy." *Warship* 41 (January 1987): 43–51.

______. "U.S. Mine-Countermeasures Programs." *International Defense Review* 17 (1984): 1259–68.

______. "World Naval Developments." U.S. Naval Institute *Proceedings* 114 (June 1988): 119–20.

Gammell, Clark M. "Naval and Maritime Events, 1 July 1966–30 June 1967." In *Naval Review* 1968, edited by Frank Uhlig, Jr., pp. 240–72. Annapolis: U.S. Naval Institute, 1968.

Garcia, Rudy C. "Yorktown, Va.—Mineman's Alma Mater." *All Hands* 481 (February 1957): 16–19.

Giusti, James R. "Sweeping the Gulf." *Surface Warfare* 13 (March-April 1988): 2–5.

Grant, Robert M. "Known Sunk German Submarine Losses, 1914–1918." U.S. Naval Institute *Proceedings* 64 (January 1938): 66–77.

Bibliography

_____. "The Use of Mines Against Submarines." U.S. Naval Institute *Proceedings* 64 (September 1938): 1275–79.

Greeley, Brendan M., Jr. "Combined U.S. Forces Defeat Iranian Mine-Laying Mission." *Aviation Week & Space Technology* 28 September 1987): 32–33.

Greer, William L., and James Bartholomew. "The Psychology of Mine Warfare." U.S. Naval Institute *Proceedings* 112 (February 1986): 58–62.

Groning, H. W. "A New Concept in Mine Countermeasures." *Naval Forces* 5 (1984): 66–72.

Haberlein, Charles R., Jr. "Damn the Torpedoes!" In *The Image of War, 1861–1865.* Vol. 6. *The End of An Era,* edited by William C. Davis, pp. 86–96. Garden City, NY: Doubleday & Co., 1984.

Hamilton, Alfred T. "Clearing the Way: LANTFLT Mine Ops." *Surface Warfare* 6 (September 1981): 28–31.

Hanks, Carlos C. "Mines of Long Ago." U.S. Naval Institute *Proceedings* 66 (November 1940): 1548–51.

Harrison, Kirby. "Danger Below: Underwater Mines and Countermeasures." *Asia-Pacific Defense Forum* 8 (Spring 1984): 35–39.

Hastings, Scott A. "A Maritime Strategy for 2038." U.S. Naval Institute *Proceedings* (July 1988): 30–35.

Heine, Kenneth A. "Sweeping Ahead." *Surface Warfare* 13 (March-April 1988): 6–9.

"This is No Drill!: Saving the 'Sammy B.'" *Surface Warfare* 13 (July-August 1988): 2–7.

Herman, William A. "Surface Minesweeps—Key MCM Players." *Surface Warfare* 7 (February 1982): 22.

Hessman, James D. "A Clear Path to Maritime Supremacy: the Fiscal and Physical Dimensions of Mine Warfare." *Sea Power* 31 (November 1988): 36–40.

_____. "In Search of 'The Weapon That Waits.'" *Sea Power* 27 (July 1984): 40–42.

_____. "Mine Warfare: A Sweeping Assessment." *Sea Power* 30 (September 1987): 7–10.

_____. "Mine Warfare: The Lessons Not Learned." *Sea Power* (October 1988): 37–39.

_____. "Mine Warfare: A Two-Sided Game." *Sea Power* 27 (August 1984): 29–32.

Hinkamp, C. N. "Pipe Sweepers." U.S. Naval Institute *Proceedings* 46 (September 1920): 1477–84.

_____. "Using An Old Wrinkle." U.S. Naval Institute *Proceedings* 46 (November 1920): 1785–88.

Hoblitzell, James J. "The Lessons of Mine Warfare." U.S. Naval Institute *Proceedings* 88 (December 1962): 32–37.

Hoffman, James A. "Market Time in the Gulf of Thailand." In *Naval Review 1968*, edited by Frank Uhlig, Jr., pp. 37–67. Annapolis: U.S. Naval Institute, 1968.

Hoffman, Roy F. "Offensive Mine Warfare: A Forgotten Strategy?" *Naval Review 1977*. U.S. Naval Institute *Proceedings* 103 (May 1977): 142–55.

Hooper, Edwin B. "The Navy Department: From a Simpler to a More Complex State." In *Evolution of the American Military Establishment Since World War II*, edited by Paul R. Schratz, pp. 50–56. Lexington, VA: George C. Marshall Research Foundation, 1978.

Horne, C. F, III. "New Role for Mine Warfare." U.S. Naval Institute *Proceedings* 108 (November 1982): 34–40.

Horsnaill, W. O. "War Beneath the Waves." *Chamber's Journal* (London), Series 7, Vol. 5 (March-May 1915): 190–92, 198–200, 293–94.

Hubbard, John H. "Mine Warfare Officers Assignments—Looking Good." *Surface Warfare* 7 (February 1982): 19.

Hughes, Hobart. "Saga of a YMS." U.S. Naval Institute *Proceedings* 74 (January 1948): 53–59.

Hutchinson, William F. "Personal Memoirs." In *Rhode Island Soldiers and Sailors Historical Society*. Vol. 8. Providence: Sidney S. Rider, 1879.

_____. "The Bay Fight: A Sketch of the Battle of Mobile Bay, August 5th, 1864." In *Personal Narratives of the Battles of the Rebellion: Being Papers Read Before the Rhode Island Soldiers and Sailors Historical Society*. Vol. 8. Providence: Sidney S. Rider, 1879.

"In the Mine Force Family." *All Hands* 481 (February 1957): 24.

"Initial Success or Failure." U.S. Naval Institute *Proceedings* 88 (December 1962): 92–105.

Johnson, Thomas M., and Raymond T. Barrett. "Mining the Strait of Hormuz." U.S. Naval Institute *Proceedings* 107 (December 1981): 83–85.

Bibliography

Jones, Mel R. "Closing the Mine Warfare Gap." U.S. Naval Institute *Proceedings* 110 (November 1984): 151–53.

______. "Mine Warfare Preparedness Begins Recovery After Years of Neglect." *Defense Systems Review* 2 (July-August 1984): 37–39.

Keller, Kenneth. "MCM *Avenger* Class—Learning the New Technology." *Surface Warfare* 12 (May-June 1987): 14–15.

Kinney, John Coddington. "Farragut at Mobile Bay." In *Battles and Leaders of the Civil War*, Vol. 4, pp. 379–400. New York: Century Co., 1884.

Kolbenschlag, George R. "Minesweeping on the Long Tao River." U.S. Naval Institute *Proceedings* 93 (June 1967): 88–102.

Lobb, K. "Mine Countermeasures." *Naval Forces* 6 (Special Supplement, 1984): 35–37.

______. "Mine Countermeasures: New Trends in Operations and Tactical Planning." *Navy International* 89 (July 1984): 421–24.

Longworth, Brian. "Mine Countermeasures." *Defence* 15 (January 1984): 7–13.

Lott, Arnold S. "Japan's Nightmare Mine Blockade." U.S. Naval Institute *Proceedings* 85 (November 1959): 39–51.

Lukin, A. P. "Secrets of Mine Warfare." U.S. Naval Institute *Proceedings* 66 (May 1940): 642–43.

Lundeberg, Philip K. "Undersea Warfare and Allied Strategy in World War I, Part I: to 1916." *The Smithsonian Journal of History* 1 (Autumn 1966): 1–30.

______. "Undersea Warfare and Allied Strategy in World War I, Part II: to 1916-1918." *The Smithsonian Journal of History* 1 (Winter 1966): 49–72.

Mack, William Paden. "As I Recall. . .". U.S. Naval Institute *Proceedings* 106 (August 1980): 105.

McDonald, Wesley. "Mine Warfare: A Pillar of Maritime Strategy." U.S. Naval Institute 111 *Proceedings* (October 1985): 46–58.

McCauley, Brian. "Operation End Sweep." U.S. Naval Institute *Proceedings* 100 (March 1974): 19–25.

McCoy, James M. "Mine Countermeasures: Who's Fooling Whom?" U.S. Naval Institute *Proceedings* 101 (July 1975): 39–43.

MacDonald, Scot. "New Surface MCM Ships." *Surface Warfare* 7 (February 1982): 24.

_____. "The Naval Action at Mobile Bay." *Surface Warfare* 12 (May-June 1987): 17–23.

McDonald, Wesley. "Mine Warfare: A Pillar of Maritime Strategy." U.S. Naval Institute *Proceedings* 111 (October 1985): 46–58.

McIlwraith, Charles G. "The Mine as a Tool of Limited War." U.S. Naval Institute *Proceedings* 93 (June 1967): 103.

MacLeod, Kenneth L. "River Warfare." In *Riverine Warfare: Vietnam*, pp. 174–79. Washington: Naval History Division, 1968,

Marolda, Edward J. "Naval Patrol Operations in the Vietnam War." *Patrol Boats 86: A Technical Symposium*. Sponsored by the American Society of Naval Engineers and the U.S. Coast Guard, March 1986.

Marriott, John. "Report on the Royal Navy Equipment Exhibition." *Armada* 11 (November-December 1987): 74–75.

_____. "A Survey of Modern Mine Warfare—Mines and Mine Counter-Measures Currently Employed." *Armada International* (September-October 1987): 38–50.

Martin, James M., and Bertrand P. Ramsay. "Seamines and the U.S. Navy." *Naval History* 3 (Summer 1989): 44–48.

_____. "Seamines and the U.S. Navy." *Naval History* 3 (Fall 1989): 52–54.

"MCM Developments in FRG Navy." *Navy International* 89 (April 1984): 212–15.

"MCM Round up." *Navy International* 93 (June 1988): 305–7.

Meacham, James A. "The Mine Countermeasures Ship." U.S. Naval Institute *Proceedings* 94 (April 1968): 128–29.

_____. "The Mine as a Tool of Limited War." U.S. Naval Institute *Proceedings* 93 (February 1967): 50–62.

_____. "Whatever Became of the Mines?" U.S. Naval Institute *Proceedings* 92 (March 1966): 115–17.

Meacham, J. S. "Four Mining Campaigns: An Historical Analysis of the Decisions of the Commanders." *Naval War College Review* 19 (June 1967): 75–129.

Mecham, Michael. "Navy Deploys Helicopters to Counter Gulf Mine Threat." *Aviation Week and Space Technology* 127 (3 August 1987): 25–26.

Bibliography

Metcalf, J. "MSH Contract Awarded—*Cardinal* Class to be SES." *Surface Warfare* 10 (January-February 1985): 13.

Milbury, C. E. "Mystery of the Magnetic Mine." *Scientific American* (March 1940). Reprinted in U.S. Naval Institute *Proceedings* 66 (May 1940): 754–56.

Miller, Richards T. "Fighting Boats of the United States." In *Naval Review 1968*, edited by Frank Uhlig, Jr., pp. 297–329. Annapolis: U.S. Naval Institute, 1968.

"Mine Countermeasures." *Naval Forces: A Special Supplement* 6 (1984): 35–37.

"Mine Countermeasures." *Navy International* 93 (June 1988): 280–86.

"Mine Countermeasures: Action Information Organisation for MCMV." *Navy International* 89 (July 1984): 404–10.

"Mine Countermeasures and the Helicopter." *International Defense Review* 6 (October 1973): 582–83.

"The Mine Menace and Countermeasures." *Maritime Defence* 14 (April 1989): 98–125.

"Mine Warfare and Countermeasures." *International Defense Review* 6 (October 1973): 577–81.

"Mine Warfare: Threat and Counter Threat." *Navy International* 87 (November 1982): 1422–28.

"Mine Warfare: The U.S. Navy Reacts." *Navy International* 90 (January 1985): 35–38.

"Mine Warfare Today and Tomorrow." *Surface Warfare* 10 (September-October 1985): 26–29.

"Mine Warfare Vessels and Systems." *International Defense Review* 17 (1984): 1654–62.

"Mines & Mining." *Navy International* 91 (February 1986): 106–11.

"Minesweeper/Hunter Design." *Navy International* 93 (June 1988): 288–98.

Moorer, Thomas H. "Mine Warfare: Entering a New Era." *Wings of Gold* 9 (Fall 1984): 7–8.

Morgan, William James. "Torpedoes in the James." *The Iron Worker* 26 (Summer 1962): 1–11.

Morganthau, Tom. "Is This Any Way to Run a Navy?" *Newsweek*, 31 August 1987, p. 16.

Mouton, E. E. "One Tour in a Mackerel Taxi—And How it Grew." U.S. Naval Institute *Proceedings* 90 (April 1964): 86–92.

Mumford, Robert E. Jr. "Jackstay: New Dimensions in Amphibious Warfare." *Naval Review 1968*. Annapolis: U.S. Naval Institute, 1968, pp. 69–87.

Mussi, Chuck. "'To See the Dawn:' The Night-long Battle to Save USS *Roberts*." *All Hands* 857 (August 1988): 4–10.

"Mystery of the Magnetic Mine." U.S. Naval Institute *Proceedings* 66 (May 1940): 754.

The Navy Department: A Brief History Until 1945. Washington: Naval Historical Foundation, 1970.

Nepean, Philip. "Naval Mine Countermeasures." *Armada International* 8 (March-April 1984): 103–21.

______. "Sea Mines." *Armada International* 9 (February 1985): 102–7.

"New Minesweeper Class." *Surface Warfare* 9 (July-August 1984): 20.

Norman, Stanley J. "Prepared For Mine Warfare?" U.S. Naval Institute *Proceedings* 109 (February 1983): 64–69.

Padgett, Harry E. "Newest Minesweepers in the U.S. Navy." U.S. Naval Institute *Proceedings* 89 (July 1963): 172–73.

Padwick, Alan. "The Development and Present-Day Use of Mine Hunting Sonars." *Military Technology* (March 1987): 66–73.

Patterson, Andrew, Jr. "Mining: A Naval Strategy." *Naval War College Review* 23 (May 1971): 52–66.

Patton, James H., Jr. "ASW: Winning the Race." U.S. Naval Institute *Proceedings* 114 (June 1988): 63–66.

Pollitt, George W. "MCM Computer for MSOs." *Surface Warfare* 7 (February 1982): 20–21.

Polmar, Norman. "The Mine as a Tool of Limited War." U.S. Naval Institute *Proceedings* 93 (June 1967): 103–5.

______. "The U.S. Navy: Mine Countermeasures." *Proceedings* 105 (February 1979): 117–19.

Bibliography

"Port Protectors." *All Hands* 481 (February 1957): 20–23.

Preston, Antony. "The Infernal Machine: Mines and Countermeasures." *Defence* 19 (August 1988): 559–66.

"Pride and Professionalism—Mine Warfare." *Surface Warfare* 7 (February 1982): 23.

Prina, L. Edgar. "Deep Threat: The Navy is Flunking Higher on Mine Warfare." *Sea Power* (May 1983): 41–48.

_____. "No Exit, No Entry: U.S. and NATO Mine Warfare Capabilities are Improving, But Not Fast Enough!" *Sea Power* 29 (February 1986): 6–12.

"Radio-Controlled Drone Boats Used in Vietnam Minesweeping." U.S. Naval Institute *Proceedings* 96 (February 1970): 123–24.

Rairden, P. W., Jr. "The Importance of Mine Warfare." U.S. Naval Institute *Proceedings* 78 (August 1952): 847–49.

Ransom, M. A. "The Little Gray Ships." U.S. Naval Institute *Proceedings* 62 (September 1936): 1280–94.

Rawson, Edward Kirk. "Admiral Farragut." *Atlantic Monthly* 69 (April 1892): 483–89.

Resing, David C. "Mine Countermeasures in Coastal Harbors: A Force Planner's Dilemma." *Naval War College Review* 40 (Spring 1987): 53–62.

Robinson, John G. "Miners Served Here." *Surface Warfare* 5 (August 1980): 26–29.

Rogers, Edward S. "Mines Wait But We Can't." U.S. Naval Institute *Proceedings* 108 (August 1982): 51–54.

"Rules of Mine Warfare." U.S. Naval Institute *Proceedings* 66 (November 1940): 1677.

Ruth, Michael S. "COOP-The Breakout Gang." *Surface Warfare* 9 (July-August 1984): 19–20.

Saar, Charles William. "Offensive Mining as a Soviet Strategy." U.S. Naval Institute *Proceedings* 90 (August 1964): 42–51.

Salitter, Michael, and Ulrich Weisser. "Shallow Water Warfare in Northern Europe." U.S. Naval Institute *Proceedings* 103 (March 1977): 36–45.

Schlim, A. J. P. "Mine Warfare in the North Sea." *Nato's Sixteen Nations* 1 (April-May 1984): 20–22.

Schreadley, R. L. "The Mine Force—Where the Fleet's Going, It's Been." U.S. Naval Institute *Proceedings* 100 (September): 26–31.

_____. "The Naval War in Vietnam, 1950–1970." *Naval Review 1971*. U.S. Naval Institute *Proceedings* 97 (May 1971): 180–209. Reprinted in *Riverine Warfare: Vietnam*, pp. 18–47. Washington: Naval History Division, 1972.

Schultz, Mort. "They Hunt For Floating Death." *Popular Mechanics*, April 1968, pp. 86–89, 214.

Searight, Murland W. "Prepare to Sweep Mines . . ." U.S. Naval Institute *Proceedings* 96 (January 1970): 54–59.

Sears, James H. "The Coast in Warfare." U.S. Naval Institute *Proceedings* 27 (September 1901): 449–527.

_____. "The Coast in Warfare. Part Two." U.S. Naval Institute *Proceedings* 27 (December 1901): 649–712.

Shelley, Marke R. "A Better Game of Dodge 'n Detonate." U.S. Naval Institute *Proceedings* 114 (February 1988): 41–43.

Sheppard, William. "Dismal Spit and Her Mackerel Taxis." U.S. Naval Institute *Proceedings* 70 (October 1944): 1253–57.

Shortley, George. "Operations Research in Wartime Naval Mining." *Operations Research Journal* 15 (January-February 1967).

Simmons, Edwin H. "Mining at Wonsan—and in the Persian Gulf." *Fortitudine* 17 (Summer 1987): 3–7.

Skripsky, Alan J. "Handling the Magnetic Force—USS *Peleliu* Depermed." *Surface Warfare* 7 (July 1982): 7–9.

Smith, Robert H. "Mine Warfare: Promise Deferred." U.S. Naval Institute *Proceedings* 106 (April 1980): 27–33.

Spooner, G. R. "Minewarfare Policy or Palsy?" *Royal United Services Institute for Defence Studies Journal* 115 (1976).

Staveley, William. "NATO's Defence Against Mines." *Naval Forces* 5 (1984): 14–21.

[Stillwell, Paul]. "SS Bridgeton: The First Convoy, Interview with Captain Frank C. Seitz, Jr., U.S. Merchant Marine." *Naval Review 1988*. U.S. Naval Institute *Proceedings* 114 (May 1988): 52–57.

Bibliography

"Submarines and Mines." U.S. Naval Institute *Proceedings* 36 (December 1910): 1192–94.

Suddath, Thomas H. "The Role of the U.S. Navy in Mine Warfare." U.S. Naval Institute *Proceedings* 91 (September 1965): 108.

Swarztrauber, S. A. "River Patrol Relearned." U.S. Naval Institute *Proceedings* 96 (May 1970): 120–57.

Tarpey, John F. "A Minestruck Navy Forgets Its History." U.S. Naval Institute *Proceedings* 114 (February 1988): 44–47.

Taylor, Jeremy D. "Mining: A Well Reasoned and Circumspect Defense." U.S. Naval Institute *Proceedings* 130 (November 1977): 40–45.

"The U.S. Navy's MCMVs—wood and glass reinforced plastic construction." *Maritime Defence* 9 (May 1984): 159–61.

Thomson, D. W. "David Bushnell and the First American Submarine." U.S. Naval Institute *Proceedings* 68 (February 1942): 176–86.

Tierney, Larry, and Richard Funk. "The Tools of the Minehunting Trade." *Sea Technology* 26 (November 1985): 47–50.

Torry, John A. H. "Minesweeping From the Air." U.S. Naval Institute *Proceedings* 87 (January 1961): 138–40.

"Troika Mine Sweeping System." *Journal of Defense and Diplomacy* 5 (1987): 41–43.

Truver, Scott C. "Mines of August: an International Whodunit." U.S. Naval Institute *Proceedings* 111 (May 1985): 95–117.

______. "Tomorrow's Fleet." U.S. Naval Institute *Proceedings* 115 (May 1989): 302–17.

______. "Weapons That Wait . . . And Wait . . ." U.S. Naval Institute *Proceedings* 114 (February 1988): 31–47.

Truver, Scott C., and Jonathan S. Thompson. "Navy Mine Countermeasures: Quo Vadis?" *Armed Forces Journal International* (April 1987): 70–74.

"US Navy's Mine Warfare Capability Improved." *Defence* 18 (December 1987): 740.

"US and UK Coastal Minehunter Developments." *Maritime Defence* 9 (January 1984): 24.

Van Nortwick, John. "Endsweep." *Marine Corps Gazette* 58 (May 1974): 29–36.

Van Nostrand, A. D. "Minesweeping." U.S. Naval Institute *Proceedings* 72 (April 1946): 505–9.

Vogel, Bertram. "The Great Strangling of Japan." U.S. Naval Institute *Proceedings* 73 (November 1947): 1305–9.

Wages, C. J. "Mines . . . The Weapons That Wait." U.S. Naval Institute *Proceedings* 88 (May 1962): 103–13.

Walsh, Don. "ROVers of the Sea." *Sea Power* 27 (July 1984): 25–32.

Walters, Brian. "ROVs—An Integral Part of MCM." *Navy International* 93 (June 1988): 299–304.

Watson, John C. "Farragut and Mobile Bay—Personal Reminiscences." U.S. Naval Institute *Proceedings* 53 (May 1927): 551–57.

Watson, Russell. "The Mines of August: the Threat from Iran Prompts a Western Buildup in and Around the Persian Gulf." *Newsweek*, 24 August 1987, p. 22.

Wettern, Desmond. "Mine Countermeasures: Forgotten Lessons." *Defense & Diplomacy* 6 (November 1988): 18–21.

White, Carl. "Move and Countermove: Belated Recognition for Naval Mine Warfare and Mine Countermeasures Requirements." *Sea Power* (June 1985): 12–22, 27–30, 35.

Wile, Ted S. "Their Mine Warfare Capability." U.S. Naval Institute *Proceedings* 108 (October 1982): 145–51.

Wilhelm, Maria. "Where are U.S. Minesweepers? Out of Service and Out-of-date, Says Arms Expert William Lind." *People Weekly* 28 (24 August 1987): 30.

Wilson, George C. "Navy is Ready to Clear Mines From Harbors of North Vietnam." U.S. Naval Institute *Proceedings* 99 (March 1973): 122–23.

Worsham, Susan. "CNO Lauds MSOs." *Surface Warfare* 7 (June 1982): 37.

Wright, C. C. "Origins of the 'Bird' Class Minesweeper Design." Appendix to Harry C. Armstrong, "The Removal of the North Sea Mine Barrage." *Warship International* 25 (June 1988): 163–67.

Index